Supernatural And Paranormal Event Case Studies

By Mark Kirby
PO Box 6517
Columbia, SC 29260-6517
All Contents Copyright © 2009

All of these stories are based on actual true events of the supernatural and paranormal. These stories are not fiction, they are not movies, and these stories are not screenplays from Hollywood. The stories in this book are real and are proof that reality can be stranger than our imagination or even stories from the mind of the most skilled writers.

All stories are based on witness interviews, investigation, photographs, or other evidence of their occurrences.

One demand by all persons interviewed or locations investigated was that their actual names or locations not be released. Several of the stories involve valuable real estate that might become unmarketable or have a lower market value if the location of the events became public. Other locations are currently in use by the public and attendance may drop for many if they fear paranormal activities.

What can be disclosed is that all parties involved are reputable persons, most with college educations or executive positions and they have no reason to fabricate stories especially since they do not want their names revealed. No persons interviewed have received or will receive any compensation for their assistance.

About the author:

Mark Kirby has been trained as an investigator and is currently a corporate investigator and investigative consultant. He has held the position of Insurance Investigator, State Investigator, and has been a licensed Private Investigator. Mr. Kirby has investigated over 1,000 accidents including aircraft and vehicle accidents and one murder for life insurance scheme that resulted in an arrest and conviction.

Mark Kirby is also an expert in computer digital photo enhancing technology and has appeared on a national radio show to discuss his findings upon analysis of UFO incident photographs involving a US Air Force fighter pilot. Mr. Kirby has also been published in international journals and publications.

Mr. Kirby has been ranked as being among the top 100 UFO investigators in the world.

CONTENTS

The Guest That Never Leaves.

When we look at old buildings we sometimes imagine all the people and events that occurred in that building.

We can imagine all the laughter, the parties, the sorrow, the anger, and many years of emotion that can sometimes become part of the building itself. You can almost feel the emotions of a building when you first walk into the building as if the building was alive and was full of memories and experiences.

In the early 1900's a hotel was built in a large city as the country was growing and cities were expanding quickly. The hotel was one of the many new types called high-rises or sky scrapers and boasted 12 floors of modern rooms built for the comfort of the guests.

On the first floor was a sprawling lobby with rich red carpet and walls decorated by ornamental plaster and wood. Rich stained wood panels and trim gave the new hotel the smell of new building. When one entered the hotel there was a mixture of the smells of new wood, paint, and food being prepared in the elegant eating area complete with large chandeliers.

Many thousands of guests spent nights, weeks, months or even years at the fine hotel.

As with everything time took its toll along with the changing needs of travelers to the city. The hotel became too expensive to maintain and was replaced by many smaller and newer hotels with such things as wireless internet and rooms set up to be used as offices by traveling business people.

Finally the time had come for the old grand hotel to shut down.

Located just a few blocks from the main campus of a large University, the hotel was perfect for temporary housing for students while an additional dorm was being built.
The hotel was saved and once again put into service to house people as it had done for so many years. The freshman students moved their belongings into the old hotel and most were excited to be staying in an historic hotel but that was to soon change. There was excitement and happiness as the students planned their school year and anticipated all the sporting events they would attend as they decorated their rooms.

The hotel dormitory was divided into two sections, the lower floors were for boys and the top floors were for girls. Each of the floors for girls was assigned an adult "House- mother" to make sure everyone was safe not only from crime, but from the boys on the floors below. Some of the young residents protested the need for a "nanny" to look over them however most of the parents of the students felt more comfortable knowing there was an adult to protect their teens as they went away from home for the first time to live on their own.

While the city was low on crime, the downtown location of the hotel in a large city made it easy prey for wandering people looking for an opportunity to steal money or worse and make a fast getaway using the downtown buildings and streets as cover.

Before the end of the first semester, there was already trouble in the temporary dormitory.

At approximately 11:00 PM a man had entered the building and was roaming the floors.

A young female student awoke to find the man standing in her room staring at her and she screamed out awakening her roommate and others on the eighth floor. The residents and the Housemother who had come to investigate saw the man in the hallway. The student described the man as angry looking with dark eyes.

The police were called and were there in minutes sealing off the building and conducting a floor-to-floor search. A description of the male intruder was given to the police. The intruder was not found and it appeared the intruder had escaped.

Additional security measures were put into place to prevent a reoccurrence of the incident along with a complete search of every room and possible hiding area in the entire building. No guests were allowed to enter before first going past the guard station and registering at the desk first. No evidence of the intruder was found after the search was completed.

Resident students of the building were warned about the incident and assured they would be safe.

During Thanksgiving holiday, the University basically shuts down and all students go home to be with their families as is the case with almost all schools.

Prior to the return of the students, a maintenance crew was performing needed plumbing repairs on the eighth floor when they spotted a man roaming the halls. The city and university police were called and officers arrived within 3 minutes and were let in by the front door security guard. The building was sealed off and searched but once again the intruder had escaped. The theory was that the intruder was ducking into the fire escape stairs and then exiting at another floor, waiting, then escaping with all the excitement.

When the students returned after their holiday they found even more security and were reminded to be alert to any intruders and to report it immediately to the security guard or campus police.

Fortunately the security measures and patrols by police had stopped the intruder from trying again as all was calm in the building, well for a month that is.

Again at about 11:00 PM a student saw a man roaming the eighth floor hallway. The police were called and the student and Housemother described the same man who had been seen previously on the eighth floor. Just as before, the man managed to elude capture.

While no one had been harmed and nothing was missing, there was evidence that someone had been moving items in the dorm rooms while the girls slept.

This situation was becoming serious and a concern of University officials and the residents to the point that the Housemother kept a baseball bat (donated by the university baseball team) with her most of the time for protection of her students.

At 9:00 PM the next week the Housemother was alert enough to see the man roaming the hallway and she chased him holding the baseball bat to use against him if needed. The man turned and looked at the Housemother and calmly went around the corner. When the Housemother rounded the corner, the man was gone and this particular corner had no exit. The

Housemother checked all the rooms in that area and the students had seen nothing and some rooms were still empty as the students were at the library or on dates.

The police were called and once again they found nothing. The Housemother was visibly upset and appeared frightened according to dorm residents and campus police.

News of the problems at the dormitory spread across the campus and the football team decided it was time for them to take care of the situation.

With the permission of the University and the House mother, volunteer members of the football team and fraternities would stand guard duty on the eighth floor in the hallway at night and they would catch the intruder once and for all. Many of the volunteers were armed with baseball bats or other blunt objects such as steel pipes to subdue the intruder and they were not kidding around. No doubt some of the students also had weapons we will not discuss.

After a full week of seeing or hearing nothing unusual the volunteers had just about decided their job was done and that the intruder had gotten word they would give him a severe beating if they found him.

At approximately 2:00 AM the next day, one of the volunteers spotted the man and gave chase calling for the others. They managed to corner the man in a dead end hallway at which point the man simply walked through the wall and disappeared. Based on statements from people present, the volunteers became extremely frightened and one even ran away.

The intruder was now described as a "ghost" and the fear on the eighth floor became even worse. The many escapes from the lockdowns now made sense to everyone. It is impossible to trap a ghost or find one if they do not want to be found.

The University School of Arts and Media was contacted and plan was made to try and photograph the ghostly intruder of the eighth floor.

As the hall lights were always on, there was no lighting problem and high-speed low light film could be used. Several time-lapse cameras were set up in the hallways to try and snap a picture of the intruder.

It took over a week however one camera did manage to capture several black and white images of a middle-aged man walking the hall with light colored shirt, dark dress pants, and dark shoes. One detail of the photograph was that the man's feet were not touching the floor and he was semi transparent in that you could see through him. There was also a type of "halo" around the man's entire body. No one else was present in the photograph.

The images were distributed around campus and were seen by many people. Some thought a creative student manipulated the image while others thought it was in fact a photograph of a real ghost that was haunting the eighth floor of the old hotel.

By this time the school year was nearing an end and a new batch of students would be living in the old hotel in the fall.

Due to the problems and complaints about the old hotel dormitory, the University managed to convince the construction company to complete the new dormitories in time for the next fall term.

The problem with the eighth floor intruder was solved, as the students no longer used the hotel for housing.

The old hotel building was located on prime real estate in the heart of the expanding city. The old hotel and property were quickly sold to a bank for a new high-rise building to house their local operations and to rent office space to other businesses. A new modern 22-story building was designed to be built in the place of the old hotel but first the old hotel had to be demolished.

The only safe way to demolish a tall building is through implosion where the explosive charges cause the building to collapse on itself in a small area. The method of imploding a building on itself is common and most people have seen it on television if not in person.

The old hotel was prepared to be demolished, the explosive charges were set in place, and the public was waiting to watch the event.

Prior to the actual demolition, the entire building has to be checked for unknown occupants or workers who might still be in the building.

The time had come, the alarm horns started blowing, and the countdown had begun. Suddenly one of the safety crew noticed a man in the window on the eighth floor and the countdown was put on hold and everything put in safety mode.

Workers rushed up the steps to the eighth floor to locate the person but no one was found there or on any other floor.

It was assumed the image was simply a reflection of someone watching from another building across the street and the countdown began again.

The horns sounded, the countdown, 3,2,1, then the explosion. The crowd let out a cheer and then a gasp. Looks of horror appeared on faces including those of the demolition workers. As the building was collapsing a man was staring out the window from the eighth floor. The crowd saw the window and face collapse into the mass of dust and debris. Witnesses stated some people screamed out in horror and several began crying over what they had just seen.

Immediately word went out that a person had been in the building when it was destroyed. The rubble was carefully searched for weeks and watched for a trace of a body as it was hauled off and no trace of human remains was ever found.

The new 22-story building was built on the site. The bank moved into their building and a short time later additional commercial tenants moved into the gleaming new building.

Within a few months people working at night on the eighth floor started calling police to tell them a stranger was roaming the halls of the building and once again each time no one was found. In fact employees were so scared of the man roaming the halls of the eighth floor at night that they refused to work on the eighth floor of that building after dark.

The building has since been sold to another company. When contacted, the company refused to discuss any issue of a person seen on the eighth floor.

I had reason to be in the downtown area of that city a year ago during the nighttime.

I made a point to go past the building to take a look at it just out of curiosity. It was not too late at night and you could see people were working after dark because the lights were on in offices and blinds were open. In fact you could see people sitting at their desks working. I did notice one strange thing, the 5th, 6th, and 7th floors had lights and so did the 9th and 10th floors but the eighth floor was totally dark.

It was a strange and eerie sight to see the building lights off on that one floor, but not unexpected.

It seems the guest has never left the building and does not intend to.

Army Of The Night

While the United States is not an old country compared to most, it has had its share of battles on its soil.

Many areas of our country are known battlegrounds for revolutionary battles and for civil war battles. Some areas have had several battles from both wars fought on them. One can only wonder about the deaths and suffering of all those people killed in battle or the hardships they experienced marching to battle.

Doug and Gary are the best of friends and have been friends for most of their lives. One reason for their friendship is that they both have the same hobby, racing and restoring classic sports cars.

Doug was a fighter pilot and the top of his class in flight school. Gary is an expert marksman and is studying another degree of black belt in Karate so the parties of this incident are not easily frightened.

One weekend while riding around looking for a new project, Gary spotted an old MGB sitting in a shed behind a house. Doug and Gary both had always like MG's of all models and in fact Gary's first car was an old MG when he was in high school.

After few visits and some negotiations with the owner, Doug purchased the MGB at a good price and with Gary's help was able to get it started and drive it to Doug's house.

Over the summer, Doug and Gary worked together to restore the MGB in hopes of taking turns driving it in the local sports car club events.

The test runs had been good and the car was declared roadworthy and dependable.

Doug, Gary, and Doug's wife would often cook out together during the summer at Doug's house. Gary's girlfriend was away visiting her parents in another state.

On this particular night Doug wanted to take the MG for a spin after consuming some grilled steaks.

Doug always kept the car in his garage and always kept the top down since he and Gary would never drive a convertible with the top up unless it was pouring rain.

It was a perfect summer night for a ride with the top down. The weather was clear and it was not too hot or too cold, about 74 degrees.

Since the night was perfect for a spin in the car, Doug and Gary decided they would drive the car out into the country and take back roads so they could do some maneuvers, test everything out, and maybe speed a little.

While they were driving, Gary noticed that the headlights were not aiming properly and suggested to Doug that they pull over and aim the headlights. Doug found a spot to pull the car off the road leaving the car running with the lights on.

Gary was guiding Doug on aiming the lights while Doug adjusted the screws on the headlight that aims the lights up or down, right or left.

As they were working, Gary heard some noises. Gary looked around but there was nothing in sight, no houses, no lights, nothing except some woods with mostly very old and tall pine trees with no lower branches and large green tops. The noises got louder and louder until Doug asked Gary " what's that all about?". Gary replied he had no idea but it sounded like a bunch of people marching with canteens, weapons, or other objects that caused a clanging metal sound along with the sound of marching feet. By the sound of the footsteps, Gary estimated the number of people marching to be at least 50 to 100 people or more.

Doug was still working on the lights but by now Gary's full attention was on the woods across the small two-lane road in front of the car. The noise was getting louder and the people were getting closer. Since there is a military base near the area, Doug and Gary guessed it was troops out on night maneuvers doing their training and Gary went back to helping Doug get the other headlight aimed properly.

A strong wind came up and Doug noted a thunderstorm must have been coming. Instinctively Gary looked up and could see nothing but stars, not a single cloud in the sky. The noise of the marching troops was loud enough to where they must have been very close but even with the car lights on, nothing could be seen except trees and the forest floor. The wind continued to increase and by now Doug and Gary could hear men talking. Both Doug and Gary thought it was very strange that they could hear so many people that close and hear them talking but not see them at all especially since there were no low branches or other objects to block their view of the wooded area. Gary commented that it might not have been such a good idea to stop in an area that had no one else around.

The wind was still blowing and the wind seemed to increase as the sound of marching people got closer. Gary commented that it was strange to have wind on a summer night and no storm especially since there was no wind earlier. Doug raised up from finishing the headlight and in a slightly frightened voice pointed out to Gary that there was no wind on their side of the road or anywhere else except for the woods where the marching sounds were coming from. Gary was part Cherokee Indian and told Doug that winds were a sign that spirits were near. Both Doug and Gary realized that the treetops where the sounds were located were blowing, the trees were bending with the wind, but it was perfectly calm everywhere else around them.

Doug threw his tools in the back of the car and told Gary if he wasn't in the car in one second he was leaving him there. Gary jumped into the car without opening the door and they both made a hasty retreat away from the area doing a short wheelie in the process.

Unfortunately Doug was so excited he was not watching the speedometer and was doing about 75 MPH as he passed through a crossroads a few miles away with a 45MPH speed limit. As luck would have it a Deputy Sheriff was waiting and caught Doug giving him a ticket which would cost Doug $ 150.00. Doug did not really have the $150.00 to give up since he had just gotten out of the Air Force and was looking for a job and his savings were going fast for house payments.

When Doug and Gary got home Doug's wife asked him why they had been gone so long to which Doug would only reply that they got lost and then he got a ticket for speeding.

Doug chose to try and get the fine reduced so he went to the Magistrate to see what could be done. The Magistrate asked Doug why he was going to so fast and Doug made up some excuse that the Magistrate did not accept. The Magistrate asked Doug the real reason why he was speeding and Doug told the Magistrate he would not believe him. The Magistrate challenged Doug to "try me" at which point Doug told the story of the marching men, noises, and wind.

The Magistrate asked Doug if it was located off a certain road and Doug replied it was. The Magistrate told Doug that he and his family had lived in that area for generations and the area was the scene of a large Revolutionary battle. Doug was also told that people who knew the area avoided it because of many strange things and that was why there were no houses, buildings, or farms around that area. The Magistrate reduced the ticket to $25.00 and suggested Doug stay away from that area at night.

I got directions from Gary and visited the site during the day to visualize the incident. The woods were exactly as described to me. With me on this investigation was an Archaeologist who studied the area and was able to find several objects from the 1700's including a deformed musket ball and a piece of a belt which verified the claim of a battle along with historical records located at the State Historical Library.

According to people interviewed, to this day people in the area will not go near those woods at night and there has been no development on or around what is still nothing but woods where strange things happen at night.

The Back Yard

Dave and Marie were very excited. Dave had just received a job promotion and they were moving to a new city not far from where they both grew up. Their daughter was not quite three years old yet so leaving school was not an issue for their child.

After some searching Dave and Marie found the perfect house in a very nice neighborhood in the suburbs only 10 minutes from Dave's job location. The house was less than 10 years old and was available for rent. As a rental house near a very large city, the real estate salesman explained that the house was often vacant as its occupants worked their way up the ladder and eventually bought a home however the house was always rented quickly and never stayed on the market long due to the very reasonable rent. The decision was not difficult. The family would rent the house and move in as soon as possible. The house was just what they wanted, it was a large tri-story with many bedrooms, and it had a large yard for the child to play. Dave and Marie had discussed the possibility of purchasing the home if they liked it after a few years and the owner gladly agreed to discuss it when they were ready.

The furniture was moved into the house, new curtains put up, and the house was now their home.

Dave's job was what he always wanted. The position was a management position with a much higher salary than his previous position and the work was what Dave loved to do. Since Dave was making good money now, Marie did not have to work and could stay at home and take care of their daughter instead of paying day care.

Neighbors would come by to welcome the family. Dave and Marie sensed the neighbors acted strange at times but they were nice and offered any help the family needed. It was always the same. The welcoming smile from the neighbors, the friendly chitchat, then a look of sadness and slight fear on their faces before they would once again smile and leave to return to their homes. Many of the neighbors had lived in the neighborhood for years since it was a great location, held its value, and was free of problems. Dave and Marie were sure they made a good choice.

The kitchen eating area had a large window facing the back yard of the home. All of the homes had fenced in back yards and the houses formed a type of square such that you could see into all the back yards from each of the houses.

One night Dave worked late so the family ate dinner later than usual. It was almost dark but not quite dark. The child's seat was facing the window and back yard. As they were eating the child said, "Man, mommy, man". The mother turned and saw no one. A few minutes later the child once again said, "man, mommy, man". This time the mother turned and saw a man; it was the neighbor behind their house going from his house to his patio. The mother explained to the child it was just the man that lives in the house behind their house.

Spring was coming and there was a patch of dead grass approximately six feet by six feet and it was very obvious since the rest of the yard was full of beautiful perfect green grass. Dave went to the local home improvement store and bought a few large pieces of grass to fill in the dead area. Dave carefully prepared the area digging up the dead grass, replacing the topsoil of that area with new topsoil, tilling the area, and then planting the grass. The yard now looked perfect. Dave watered the area until the new grass was growing and all was well.

Dave looked out at the back yard after getting home from work and admired what looked like a well-maintained perfect golf course. Since the next day was Saturday, Dave decided he would mow the grass the next morning. When Dave pulled the lawnmower out, he noticed the area of new grass was dead. The area looked as if it had been dead for a long time just like it did before he fixed it almost a month earlier. It was odd, the grass was fine the night before and died overnight and looked like it had been burned. Dave was upset but mowed the grass. Even though the new grass area was dead, it was tall enough to need mowing since it had been growing. As Dave pushed the mower over the new area, the mower shut off. Dave must have forgotten to fill the mower with gasoline. Dave went to the storage shed and carried the gasoline can to the mower. When Dave removed the gasoline cap he saw the mower was half full. Dave replaced the cap and tried to start the mower and it would not start. The mower was pushed into a shaded area under a small Dogwood tree and Dave took a break. After a cool drink and some rest, Dave tried to start the mower and it started and ran fine. Dave started mowing again and once again when he tried to mow the new grass area the mower shut off. Dave had enough by now and gave up going inside to watch television, as he would deal with it another day. After rolling the mower to the storage shed Dave gave the rope a pull just to see if the mower would start. The mower started on the first pull and ran fine. A puzzled Dave put the mower away for its next use.

Dave had some suspicion that one of the neighbors had thrown some type of substance on the area and killed the grass but he didn't know why because he had done nothing to upset any of the neighbors and they did not seem like the type to vandalize is yard.

Frustrated by the grass situation, Dave decided he would plant a nice medium sized ornamental tree in the dead grass area since there were no trees in that part of the back yard and they had just about decided to purchase the house the following year.

Dave chose a healthy Bradford Pear tree and since it was still spring, the tree would take root and grow fine. Once again Dave prepared the soil for a new planting. The dead grass was removed, a hole was dug, and the tree was planted. Within a few weeks new bright green sprouts were coming from the tree in that it has rained a lot during that time. The tree was healthy and growing well the problem was solved.

One evening after working on papers for his office, Dave fell asleep on the couch in the den. Since the house was a tri-level, the den was the bottom room, which opened to the back yard through a sliding glass door. Dave was sound asleep when he felt something on his face like a hand. Half asleep, Dave reasoned it was his daughter checking on him.
Dave woke up and no one was there, he told himself he had dreamed it all.

On another occasion the family was in the den watching television when Dave noticed a large spider on the wall that he quickly swatted with a magazine and disposed of. An hour later Dave noticed two more spiders on the wall and also killed and disposed of them. After another half hour three spiders appeared and Dave killed them. By this time Dave's wife and child did not want to be in the den until it was sprayed for insects.

Dave's wife called the exterminator who gave the entire house a good spraying. The exterminator had checked the foundation and areas of the house but found no Spider nest or any other spiders.

Dave had once again fallen asleep on the couch watching television late one evening when he was awakened by a sting on his face. Dave instinctively swatted that area with his hand. When Dave jumped up and looked he saw a large black spider dead on the couch apparently from the blow. Next to the couch were a dozen black spiders which Doug frantically killed with his shoe. Dave's face was bleeding so he cleaned the area and put antibiotic ointment on the wound.

Another call to the exterminator and another treatment of the house was apparently needed and done. The exterminator told Dave and Marie it was very unusual for him to treat a house and then have insect pests present so soon after the treatment and that neighborhood had never had a spider problem that he knew of since there were very few large trees and no cool places for the spiders to make nests. Doug was also told by the exterminator that he had found no evidence of any spider activity such as webs, egg nests, etc. The exterminator also explained to Doug that since the house was on a concrete slab, there was no area under the house for them to nest and the bricks were sealed very tightly to the slab and the attic was free of any traces of insects or pests. Since the windows were fairly new, had screens, and were kept closed, the exterminator was at a loss to determine how the spiders were getting into the house especially so many mature spiders.

Doug had to leave on a business trip and Marie had to stay alone in the house for one night.

Marie checked all the doors, made sure they were locked, put the child to bed and went to sleep. At about three o'clock in the morning Marie heard what sounded like the child walking around in the upstairs bedroom area of the house. Marie went to check and found her daughter sound asleep in her bed. Marie rationalized it was just some noise from the street or maybe she dreamed it all.

Doug returned home the next day and Marie told him the house made noises and they both laughed.

That night both Doug and Marie were awakened by what sounded like a person pacing around in the empty bedroom at the end of the hall. Doug grabbed a flashlight and slowly went to the room suspecting an intruder was in the house. When Doug got to the bedroom he saw nothing, the windows were shut tight, and since it was the top floor of the house an intruder would not be likely enter there any way. After checking on his daughter Doug went back to sleep.

In the first week of November the family was sitting in the eating area eating Pizza they had delivered a few minutes earlier. The time was six thirty and the sun was just about to set. Once again the child looked out of the large window facing the back yard and said "mommy, the man is back". Both parents turned around that thought they saw someone. Doug went to the window to look again but saw nothing.

The next day was Saturday so the family slept a little late.

As he was making breakfast Doug looked out the window to the back yard and noticed his tree was dead. Doug rushed out to the backyard and inspected the tree that looked like it had been dead for years. Since the tree was fine and the leaves were changing colors, Doug was sure a neighbor had vandalized the tree with some type of herbicide.

Doug asked his neighbors one by one if they saw anyone near the tree of if they saw anyone pouring a substance around the tree. All of the neighbors said no however one neighbor did mention that they had lived behind Doug's house for several years and nothing would ever grow in that area. The neighbor explained to Doug that previous occupants had also tried to grow grass or shrubs in that area and they had all died too. The neighbor suggested the soil must be over a substance that leaches out and kills whatever is planted in the area.

It was after their Thanksgiving meal and Doug was watching a night bowl game on the television when he fell asleep on the couch. Doug was awakened by a stomping and pacing noise in the upper areas of the house that went on for at least fifteen minutes. Doug went to check on his wife thinking she was upset about something or was ill. Doug found his wife asleep in bed and his daughter was also asleep in her bed. Doug went to bed and fell asleep only to be awakened by what sounded like a single gunshot. No dogs were barking, no neighbors were up and no lights were on in neighborhood houses, so Doug realized he dreamed the gunshot.

On one of his morning walks, Doug saw a neighbor walking to his door. Doug spoke to the neighbor and asked if he had minute. The neighbor was very friendly and willing to talk to Doug. Doug asked the neighbor how long they had lived on the street and the neighbor informed Doug they were one of the first couples to buy a house in the new development and had lived there ever since. Doug then asked the neighbor if he had ever heard of anything strange about the house he was renting.

The neighbor told Doug that the house was constantly being rented to new people. It seemed renters would not stay in the house for more than a year or two and even after many attempts to sell the house and the owner could never sell the house. The neighbor also told Doug that previous renters had told him about problems such as large spiders and noises in the house at night.

Doug became very interested when the neighbor asked him if he heard the history of the house to which Doug replied he had not.

The neighbor told Doug that a man and his wife had bought the house new and moved in. After a couple of years the man began drinking heavily and having regular loud arguments and fights with his wife. The arguments got so bad that several times neighbors called the police. After many fights, the wife left the man who stayed in the house. The man's drinking problem got worse and he lost his job. The neighbors would notice upstairs lights on at all hours and could sometimes see the man pacing around in the back bedroom acting disturbed.

During a Thanksgiving week, the man had been drinking and was very depressed. The intoxicated and distraught man went outside to the back yard. The neighbors heard a gunshot and rushed to find the man dead lying on the ground. The authorities were called, investigated the matter, and ruled it suicide.

Doug asked the neighbor where the man died and the neighbor told him to look for an area of the yard where nothing would grow.

When the year's lease was up, Doug and his family left the house and purchased a new home on the other side of town wanting to get as far away from the rental house as possible.

The house is still rental property and cannot be sold, and to this day renters do not stay more than a year before leaving in fact some have left suddenly breaking the lease in the process. If you look in the back yard you can still see a dead area where nothing will grow.

I was able to visit the house. When I entered the house it was occupied, but the house still had an unpleasant feeling of emptiness and the feeling of a restless spirit. The occupants of the house informed me they had heard noises, seen a man in the back yard several times and had a problem with large black spiders. The residents also informed me that the house gave them a bad feeling and sometimes smelled like alcohol even though neither of them drank alcohol.

I suggested to the residents that they might wish to speak with the neighbor four houses down on the right and that is all I told them.

Tunnel To The Unknown

Of all the stories in the book this one is the most bizarre. It is hard to believe it actually happened but based on witness accounts and physical evidence it did in fact happen just as the witnesses said.

The time had come for Sam and Melanie to move from their small rental house in the city and hopefully purchase a larger house.

Sam had grown up in the country and had convinced Melanie that living in the country was the best life for them and their children.

A friend of Sam told him he had seen an old house for sale just a few miles from where Sam grew up and although it had been vacant for a long time the house was in good shape. The friend and Sam discussed how Sam could probably rent the house with an option to buy it later with the rent serving as down payment.

Sam was very excited about the house and the possibility of being able to buy the house later. Sam's friend provided him with the name of the Real Estate Company that was selling the house and directions to the house.

Before Sam called the Real Estate Company, he wanted to look at the house.

The house was not located off any main road and was overgrown with bushes and vines because it had been vacant for so long. Sam and Melanie carefully looked at the exterior of the house and determined it was still in good shape except for some painting and clearing away of overgrown shrubs and small trees that would be needed. Looking inside through the windows the couple liked what they saw as the house was very old with beautiful wood molding, exposed beams, and several large fireplaces including a very large fireplace in what was likely the study.

Sam called the Real Estate agent and arranged a tour of the inside of the house. The house was well built as Sam had expected and did not show any signs of being vacant for years except for dust and a few spider webs and the usual things you find in an old vacant house.

The Real Estate agent told Sam that a man who had inherited large parcels of land from his grandfather built the house in the late 1800's. According to the agent, the land had been in that family for many generations. The agent also told Sam that there was a much larger plantation house that had burned down during the civil war and that there was a slave graveyard on the property that was visible because many of the graves had sunken in over time. Over time the land had been divided between family members as people died so by now the original plantation no longer existed.

The house Sam was looking at was located on the same property as the slave houses which were long gone by now. In all, the house was sitting on twelve acres of land some of which was still suitable for farming. The house itself was a one story wood frame house just under eighteen hundred square feet in size. The walls were wood or plaster, and the floors were all southern pine and had a gloss from the many years of being cleaned and polished. The roof of the house was metal, and the main foundation was built from local stones. Several large oak and pecan trees surrounded the house along with old growth pines.

Sam liked the house and after several more visits he and Melanie determined it would be perfect for them. The owner was an elderly man and agreed to rent with option to buy with the rent going towards deposit. The owner reasoned he would not be around much longer due to his age. At least the owner would have someone taking care of the house and some extra income. The owner had never lived in that particular house himself and had inherited it from his Grandmother many years ago.

Sam and some friends spent an entire weekend clearing the brush, landscaping, painting, and getting the house and property suitable for the day they would move in. Melanie had many friends who helped her clean the inside of the house. By the first of the next month Sam and Melanie were ready to move their family to the house that by now was very attractive.

Moving day came and was a sample of things that would follow. The moving truck broke a suspension part on the dirt road leading to the driveway and the move was delayed until another truck arrived and the contents were shifted from one truck to the other.

When it came time to move the items into the house, an older African American gentleman named George who had worked for the moving company many years, took one step into the house and backed out refusing to go in again. The supervisor asked George what the problem was and he stated he could sense "bad spirits" in the house and he was not going back in. The other workers laughed and the supervisor told George he could help move things from the truck to the ground if the house bothered him so much.

After all of the items were moved into the house, Sam was going to his car to get some small items when George approached him and said "mister, you are making a big mistake, you don't want to live in that house, I sense suffering and death". Sam was surprised and a little shocked but thanked George for the concern that was clearly visible on George's face. George said he would pray for the family. Sam never told his family about George's concerns, as he did not want to frighten his wife and children over someone's superstition.

Along with the three children, Sam and Melanie had two male dogs. One dog was a large retriever that hunted with Sam and the other dog was a miniature poodle that spent a lot of time in Melanie's lap.

Sam wanted the dogs to get used to their new house. Sam went inside the house and called his retriever who came to the front door and then stopped and appeared afraid to come into the house. After many attempts the dog would not enter the house. At one point Sam tried to drag the eighty pound dog into the house by his collar but the dog pulled back and refused to enter, Sam gave up on introducing the retriever to the inside of the house. It was not really important that the retriever go into the house since the retriever was an outside dog and was always happier living in a covered pen in their previous yard as is the way with most outdoor hunting dogs. The poodle had no problem going inside the house and made itself at home on the couch in the study.

There was one problem with the poodle regarding the house. The rear entry to the house was at the end of a very old looking stone path leading to the back porch and back door.

Each time the poodle would use the path to the back door or was carried on the path in Melanie's arms, he would start barking at a particular spot on the side of the path approximately three feet from the side of the path and eight feet from the back porch. The barking situation became so annoying that the poodle was no longer put out of the house or brought into the house using the back door. Sometimes the poodle would end up on the path again after getting loose and barking at the spot as if something was there but there was never anything to bark at when Melanie would rush to see what the barking was about. The only thing Melanie noticed was a slightly sunken area in the ground near the path.

The retriever would not bark near the path as the retriever refused to go near the path even if he was enticed by a leftover steak bone or table scraps. The family thought the situation with the stone path and their two dogs was strange but it did not really bother the family as they loved living in the house and having so much land to explore and enjoy.

A few months after they had moved into the house, Melanie saw Sam moving around the back of the house. The children were playing outside with the poodle. As she was walking through the study, Melanie called Sam's name to help her move the kitchen table so she could clean under it. As Melanie called Sam's name towards the back of the house, Sam's head raised up from the sofa in the study where he had fallen asleep watching a football game.

Melanie was totally shocked to see Sam on the sofa as she had just seen him in the back of the house. Melanie told Sam she thought a stranger was in the house. Sam grabbed a twelve-gauge shotgun from the wall, loaded it quickly with 00 buck shot, and headed for the back of the house. Sam searched every room and closet including under the beds. No person was found. Sam told Melanie she had an over active imagination.

On another occasion Melanie heard Sam call her voice from the back bedroom. Melanie went to the bedroom only to find Sam was not there. Melanie found Sam shortly afterwards walking towards the house from the woods with their retriever. When asked how long he had been outside, Sam said he and the dog had taken a thirty-minute walk on the property and he was just getting back.

Incidents of Melanie hearing or seeing someone in the house continued however neither Sam nor the children saw anyone or anything.

Four months after moving into the house, Sam was watching television in the study and saw movement of a shadow behind him. Sam asked Melanie what she wanted. Melanie responded that she was cooking in the kitchen and told Sam she was not in the study earlier. Sam told Melanie that her imagination was starting to infect him too.

Not long after the last incident, Melanie and Sam were awakened at approximately 2:00 AM by a man's voice in the study area. As a long time hunter and living so far away from other people and houses, Sam always kept a loaded 9 mm pistol locked in a gun safe next to their bed for protection. Sam quickly entered the combination to the safe and had the pistol in his hand as he headed toward the study after checking to make sure the children were in their rooms and safe.

As Sam entered the study area it was slightly illuminated by the moon and open drapes in the room. Sam could see a dark figure of a man standing in the corner staring at him. Sam pointed the pistol at the figure and told him not to move or he would shoot. Sam reached over and turned on the light switch illuminating the room. There was no one in the room and by this time Sam was starting to get worried about the situation and wondered if somehow the man had escaped because Sam had hesitated to shoot for fear he might shoot a family member or innocent person such as a deputy or fireman who was responding to a 911 call about the house.

Sam and Melanie decided not to tell the children about the stranger but did tell the children if they saw a strange man near the house to make sure the windows and doors were locked and to let them know immediately.

Sam and Melanie had an alarm system installed in the house but the sightings of the man continued along with Sam or Melanie being touched by someone who was not there, cold wind in the house, and a bad odor that could not be located.

It had now been almost six months since Sam and Melanie moved into the house.

On a Thursday night Sam and Melanie had gone to bed around 11:00 PM and had fallen asleep. At approximately 3:00AM, Sam was awakened by Melanie holding on to the headboard of her bed screaming "don't let them take me, don't let them take me!" Sam felt like he was about to have a heart attack he was so shocked. Sam grabbed Melanie and held her asking what was wrong.

Melanie explained to Sam that she had awakened a few minutes earlier and when she opened her eyes she saw a tunnel where the outside wall to the bedroom should have been. Inside the tunnel were dozens of people in what looked like a dark cave and they were reaching out to her trying to drag her into the tunnel.

Sam told Melanie everything was fine and it was just a nightmare. Sam turned on the light next to the bed to show her the wall to the bedroom was intact. Melanie was so terrified she could go back to sleep that night.

The same event continued at least once a week and each time Melanie would end up closer to the foot of the bed towards the outside wall before she could scream out for Sam. Sam and Melanie started sleeping in the study on the sofa to Melanie would not be bothered by the people and the cave. The problem was the dark man in the corner would appear and each time Sam would have to get up and check the whole house for an intruder.

By now the retriever would not come near the house and would put his tail between his legs and put his head towards the ground if he looked in the direction of the house. The retriever had lost all interest in playing or hunting.

On a Tuesday afternoon, Melanie had gone out the front door with the poodle to get the mail. While Melanie was walking towards the mailbox at the end of the driveway the poodle ran off to the back of the house. Melanie heard the poodle barking repeatedly and then heard the dog yelp and go silent.

Melanie rushed to the back of the house and found the poodle motionless laying in the sunken area next to the old stone path. Melanie grabbed the dog and rushed to the nearest veterinarian that was 10 miles from their house. The veterinarian examined the dog and told Melanie it was dead. Melanie requested an autopsy on the dog and the results showed the dog had died of a type of heart attack. The poodle was only four years old. The doctor explained that like people dogs could have sudden heart failure for no reason but Melanie did not buy that explanation. Melanie was sure something had scared the dog to death.

Melanie called Sam at work and he came home as quickly as he could. The children were still in school so they were not aware the poodle was dead.

Sam and Melanie decided the dog's death was the final event that would force them leave the house as quickly as possible. Sam called the moving company and had their belongings put in storage while the family lived with Melanie's family in a large home in the city until they could find another house. The owner of the house made no effort to enforce the lease and remained silent on the issue of the contract.

Sam and Melanie purchased a new home near Sam's workplace and have experienced no unusual events in almost three years since they left the old house.

The old house still sits vacant on the property but has been unoccupied since Sam, Melanie and their family lived there. The house is overgrown to the point that you would not see it unless you knew it was there and the house appears to be falling in due to termites and lack of maintenance. The property and house are not for sale.

Rude Jumpsuit Man

Across the United States there are huge numbers of old buildings that have been converted from their original use to retail or office spaces.

One example of the converted buildings are the many factories built during the late 1800's and the early 1900's. Those buildings were well built and have stood the test of time. In addition to being sturdy, older factory buildings have open desirable spaces for other uses or companies. The old factory buildings are usually located in an urban area and offer large amounts of square footage not available to newer buildings in downtown or urban areas. Fortunately many of those old factory buildings have not been torn down. Unfortunately, many of those old buildings have stories and unexplained events due to their age and history.

One old converted factory building sits in the most desirable area of a medium sized growing city. The factory that was located in the building for almost one hundred years produced the same products. The factory employed approximately three hundred people but was closed not so long ago when the production was moved to a foreign country as so many U.S. companies do.

The four-story factory building is of beautiful design, sturdy, and with its open floor plan is perfect for any business.

Once the factory was closed, the building was sold and the interior was remodeled but much of the character and charm of the building was maintained. The interior of the building was left open with small spaces being dedicated to office and administrative space and storage areas for office supplies etc. The building is occupied by employees during the day and is visited by many people each year due the nature of the business located in the building.

Like all businesses, an employee is eventually going to do something to offend or upset a client or visitor and this business was no exception.

The first complaint received by the manager from a visitor was only a few months after the business moved into the renovated building.

The manager was contacted by the visitor and informed that one of the employees was extremely rude. When asked what happened the visitor told the manager that she had been walking toward the elevator (factory style freight elevators still in use) and had asked the employee to hold the elevator. The visitor stated the employee gave her a "dirty look", closed the door, and the elevator went to an upper floor.

The manager apologized and said she would talk to the employee but needed a description.

The visitor told the manager it was a man, about thirty, with dark hair and dark eyes wearing a dirty gray jumpsuit.

The manager informed the visitor that no employee fit that description and it must have been someone who wandered in off the street or possibly a member of the crew working on the air conditioning system. The visitor thanked the manager for her attention to the matter and left.

When the manager saw the supervisor for the air conditioning contractor she explained what had happened. The supervisor told the manager that all of his employees wore the same uniforms with navy blue pants and white work shirts with patches and logos and that none of them had worn any type of overalls or jumpsuit at the building.

After thinking about the situation, the manager emailed everyone in the building to be on the look out for an unauthorized man roaming the building and gave a description. No employee had seen such a man. Before closing, the manager asked co-workers to join her in searching the building before they left for the day. A search would be easy since a large part of the building was open and the offices were occupied all day so no one could have hidden in them. The workers searched each of the floors and found no one or anything usual.

Two weeks after the first sighting of the man another visitor came to complain to the manager about a rude employee. A description of the rude employee was given and it was identical to the first one given by the other visitor. This time the man in the jumpsuit had ridden from the first to the third floor on the freight elevator with the visitor. The visitor tried to make conversation with the man but the man ignored him and would not even look at him. The visitor got off at the third floor and the man continued up which would be the top floor.

Once again the manager apologized and explained that someone was apparently coming into the building and roaming during the day and that the man was not an employee of hers.

After the complaining visitor left, the manager called the police department and explained the situation. The police officer said the man was probably wandering in to look for purses or other items that could be stolen, as there had been some complaints from other businesses in the area regarding that situation. The police department promised to be on the lookout for man fitting the description in the area.

Later for the first time the man in the jumpsuit was spotted near the elevator on the ground floor by a male employee. The employee shouted at the man and ordered him not to move. The man ignored the employee and calmly entered the waiting elevator through the open door. The man in the jumpsuit looked straight ahead at the male employee and had a blank stair on his face. Before the employee could get to the elevator the door closed and the elevator went down to the basement area. The basement was used for minor storage and held electrical circuits; water pipes, sewer pipes, and other building mechanicals but had no interior walls or obstructions other than large structural support beams.

The employee immediately reported the incident to the manager who called the police department. Within five minutes several police cars arrived and the officers searched the basement, the building and the surrounding area but found nothing. The search of the basement was not difficult as there was really no place for an intruder to hide from view. A check of employees found none of their belongings were missing.

By now reports of the continuous incidents had reached the main offices located in another larger building. The top management decided to hire a private security firm to try and catch the intruder and see if he was present after the building closed and the employees left.

The guard night guard sent by the security company the first night had recently retired from the Army as decorated combat veteran after almost thirty years of service and was working to supplement his retirement.

The guard's report showed that the next evening when he reported to the building he had seen a man in a jumpsuit in the building and that he had chased the man. The man managed to get out of the building before he could catch him. The police were notified by the security company of the event. The guard did mention to his supervisor and in the report that the man in the jumpsuit acted as if he was deaf or possibly blind in that he never looked at the guard and did not seem to hear his commands.

A week later the same security guard saw the man again and chased him only to once again have the man escape this time by getting on the elevator just before the doors closed and riding the elevator to the basement. The basement was searched and all exits were immediately locked but no person was found.

After several more incidents over the next six weeks, the security company contacted the manager and informed her that they would no longer be able to provide services to that location. No further explanation was given by the security company for their refusal to provide additional security services but when pressed by the manager, the security supervisor said the original security guard and his co-workers had refused to work at the building and that was all he was going to say about the matter.

Upper management decided not to attempt to contract with another security company but they did require the manager to report to the main office to explain what the problem was with the situation in the building and why she could not handle it. The manager explained she had done everything possible but had no explanation.

As was required by regulatory officials, the elevators were inspected yearly for defects of safety issues.

The elevator safety inspector arrived to perform the safety inspection.

After what seemed like a longer than usual time period, the inspector notified the manager that he would have to shut down the main elevator as a part was worn and was a serious safety issue.

The elevator inspector also advised the manager that there was a large stain on the inside of the elevator shaft just below the second floor and it appeared as if someone had thrown a bucket of some type of chemical on the shaft. While the stain did not represent a safety hazard in his opinion, the inspector notified the manager of the stain in case someone had vandalized the building.

As required, the elevator company was contacted and the old elevator was returned to proper working order with a new part.

A few days later a gentleman was in the old building on business and mentioned to the manager that he had worked for the company that owned the building before they shut down the factory. He also told the manager that his brother, his father, and his grandfather had worked for the factory over the years and how sad he was when they closed it down.

During the usual polite conversation the gentleman asked the manager if anyone had seen the man in the jumpsuit. The question took the manager by surprise but she was relieved that the problem started before she took over.

The manager asked the gentleman what he meant and the man proceeded to tell her how there was a man in a jumpsuit that would roam the building but they could never catch him. The gentleman also told the manager how the jumpsuit man was well known by employees of the factory including his grandfather. The Manager told the man that was not likely since the man they had seen was too young to have been roaming the building that long ago.

The gentleman proceeded to tell the story handed down from generations of workers about the jumpsuit man. The jumpsuit man was a maintenance man who worked for the factory in the 1920's. One of his duties was to keep the freight elevators in working order. The main elevator that most people used began stopping and starting without warning. The jumpsuit man was called asked to fix the problem. As required, the jumpsuit man had cut off all safety switches while he inspected the elevator. As the jumpsuit man was leaning into the shaft from the open second floor doors a short circuit allowed the elevator to go down from where it was stuck just above the second floor doors. When the elevator started down it decapitated the man with blood pouring down the elevator shaft wall. The factory was shut down for a day and most workers attended his funeral.

According to the gentleman the jumpsuit man only appears when there is a safety problem with the elevator and then goes away again when the problem is fixed. In fact the factory had a policy of any time the jumpsuit man was seen there would be an immediate inspection of all elevators in the building.

As of this day the company in the building has the elevators inspected twice a year every year and they have not seen the jumpsuit man.

The same company is still in the building and has in fact built new buildings on the property to expand.

Company policy actually states that no employee is to discuss anything related to a man in a jumpsuit or "ghosts".

This story is unusual and rare in that jumpsuit man was not only seen from a distance by witnesses but people had actually ridden in the elevator with him and seen him standing next to them.

Jumpsuit man's appearance was so real as to make people standing next to him to believe he was like anyone else. If anyone did not believe in supernatural events, they would after riding in the same elevator as jumpsuit man after reading this story.

Friends in death

Tom and Mike were very close friends as they had a lot in common.

Tom had never met Mike until he started dating Mike's sister. After a long relationship Tom married Mike's sister and by then Mike and Tom were best of friends.

Tom and Mike spent a lot of time together boating, fishing, and hunting. When possible they would play basketball with other friends or throw football at a nearby empty football field.

Tom worked as a flight instructor at the local airport while Mike was still in college working on his master's degree in engineering.

Tom was not making as much money as he had hoped offering flying lessons so when a wealthy business owner offered him a job as a corporate pilot Tom could not have been happier.

Mike completed his education and began working for a small company designing and developing new products and improving current products.

Mike and Tom had been close friends for over ten years by now.

Even though Tom and Mike's sister had moved to another state, they all still visited each other and kept in touch. The two men were more like brothers than friends. The relationship between Mike and Tom was one case where the brother in laws actually did get along well especially since Tom was so good to Mike's sister.

The company Tom worked for decided it was time to purchase a faster and newer aircraft so Tom was sent to be trained at the factory prior to the purchase. Tom was an excellent pilot and the instructor at the factory told the CEO of the company that Tom was one of the best pilots he had ever flown with, and also the most careful.

Tom and Mike had flown together many times with Tom instructing Mike on how to fly aircraft. After Tom moved Mike continued the flight instructions at the local airport with another pilot and was able to obtain his private pilot's license. Mike's flight experience was limited to small single engine aircraft and clear weather as he was not yet instrument rated.

Mike and Tom kept in touch by telephone talking a couple of times a week and Mike could check on how his sister was doing in their new apartment. Mike's sister was looking forward to starting her new job as a nurse at the local hospital.

During one conversation Tom told Mike that there had been problems with the new corporate jet. Since Mike was an engineer and now a pilot, Tom wanted his opinion.

It seemed one of the two engines on the jet was having some problems and the plane had experienced vibration and control problems during several flights. Each time he would experience the problem, Tom would have the plane inspected by an aircraft mechanic. After several attempts to find or repair the problem Tom flew the plane to the factory where they determined the cause was in the engine and most likely an out of balance turbine blade. Mike asked Tom if they replaced the engine and Tom said they had. Mike then asked Tom if the problem was solved and Tom said he had not had another problem since they replaced the engine. Mike had a bad feeling about the situation and said so to Tom. Tom told Mike he worried too much but in his voice Mike could hear concern over the statement he had made especially since Mike was an engineer.

Tom invited Mike to come visit and they would take a flight on the corporate jet together. Mike liked the idea but at the time he was not able to leave town due to his job and other matters. Tom told Mike that if he did not come visit and fly on the jet they might never get another chance.

Deep down Mike had a bad feeling about the aircraft Tom was flying and had serious doubts as to whether or not the problem was actually fixed. Something told Mike there was a bad problem with the jet but he did not push the issue with Tom. Mike was always one to worry more than he had to.

The following week Tom was scheduled to pick up some executives and fly them to a corporate meeting in the next state. Mike was scheduled to fly to the same city by airline later during the same week for a conference and the two planned to get together and tour the town.

Early Monday morning the corporate jet with Tom as the pilot took off with another pilot and several corporate executives on board headed for their destination which would only take a couple of hours on the fast corporate jet.

Mike was planning to fly a small aircraft that night since the weather was clear and he was practicing flying at night and was working on his instrument rating. There was a small chance of thunderstorms but they were minimal.

Mike had dinner and was sitting in front of the television watching a re-run of a classic television comedy series that came on every night at 6:00. While he was watching television at around 6:25 Mike started feeling like he was going to die. After not more a minute Mike's entire body was shaking like someone had put him in a blender while at the same time Mike felt horror and fear like he had never felt before. Mike thought he was dying of a stroke or similar event. After a total of three minutes, Mike's symptoms stopped and he felt better. Mike was still worried and a bit confused and had a feeling he did not know where he was but after an hour he was fine.

Mike thought about not flying that night but everything seemed fine and he really wanted the flight hours. Mike thought back on what he ate and he remembered someone in the family had an allergy to shrimp and since Mike had eaten shrimp he figured out it was a reaction to the food that caused his scare. A quiet flight at night was just what Mike needed to calm down because he was still having a strange type of anxiety attack.

Mike took off solo in a small single engine aircraft at 8:20 P.M. with clear skies. Off in the far distance Mike noticed some lighting strikes but they were in the opposite direction of his flight plan and were of no concern to him. The winds were still mild and aside from the usual thermals that caused the small plane to rise or fall suddenly, the flight was smooth. There was an issue the previous week with the dipstick leaking oil on the engine but the flight mechanic had ordered a new part and replaced the old one.

Mike's parents were asleep when the phone rang at 1:12 A.M. When Mike's father said hello he knew something terrible was wrong. Both Mike's father and mother were in shock upon learning that there had been a plane crash and the pilot was killed. The first thing Mike's father did after hanging up was to try and call Mike's number to see if Mike would answer his phone. The phone rang over and over but no one answered. Just before Mike's father was going to hang up someone answered the phone. Mike had answered the phone after finally waking up.

Mike's father explained to him that a friend of Tom's had called and that Tom's plane had crashed and that he and everyone on board had been killed. Mike's father told Tom to come pick them up and they would drive to his sister's house so she would have family with her when they notified her of the plane crash. Six hours later Mike, his father, and his mother showed up at his sister's house. Mike's sister knew something horrible had happened because it was early in the morning and with her family was a deputy and several co-workers of Tom.

The next afternoon after everyone was in less of a state of shock, the coworkers told the family what had happened.

The executives had met some friends at the meeting and invited them for a flight to a nearby resort on the new jet and a return later that night. The plane took off at 5:00 P.M. for its destination and climbed to 20,000 feet when Tom noticed severe vibration near the same engine that had been a problem before. Tom radioed a mayday and told the controller the plane had severe vibration and was breaking apart. Not long after that radio transmission other aircraft in the area transmitted that they noticed a huge red flash in the sky with a flaming trail like a meteor. Tom's plane had broken up and most of the larger pieces landed in a remote swamp. It had taken rescue workers hours to get to the crash scene but there was no hope of any survivors.

According to the accident investigation, Tom's plane started to break up at 5:25 P.M. and exploded at 5:28 P.M. The cause of the breakup was failure of the rear flight control system due to a defect.

Mike only realized what had happened when he heard the time of the incident. The time was exactly the same time that Mike had experienced the feelings of terror, vibrations, shock and that he was going to die.

Even though Mike experienced the feelings of the event at 6:25 and the event started at 5:25 that was not an issue. The area where Tom crashed was in a different time zone west of Mike and exactly one hour earlier than where Mike lived so the time was actually identical.

It is not that unusual to find that people have sensed something was wrong or that something bad was going to happen. What is unusual with this event is that another party actually felt the death of someone exactly as it happened. While many people wonder what it was like when a friend of loved one died in an accident, Mike does not have to wonder as he knows exactly what it feels like to die. Mike wishes he could forget the entire incident especially the feeling of dying but he cannot.

Mike has never flown again since that terrible night and does not plan to fly again. If Mike needs to travel a long distance he drives or takes a train. If Mike were to have to go to Europe he would take a cruise ship or not go.

Both Tom and Mike know exactly what it feels like to actually die in a plane crash and Mike does not want to risk dying like that again.

Critter invasion

A family rented a vacation home on the coast for summer vacation. The family packed their favorite clothes in the car and took along their dog rather than board it at a local kennel.

The house was located near the ocean but not directly on the ocean.

Like most new houses on the coast, that particular house was on stilts to protect against flooding or hurricanes. The house was located on a marsh and was surrounded by a mixture of marsh areas and dense coastal woodlands. There were other scattered houses on the land that would allow houses to be built.

Due to the house being on stilts, there was a large area under the house that was perfect for pets. The owners of the house had fenced in the area under the house and installed a locking gate to secure peoples' dogs. Also under the house in the fenced area was a large custom-built doghouse to accommodate even large dogs.

The family dog was a gentle German shepherd and was excited to have the room to roam and explore and she immediately claimed the doghouse as her own.

Since the house was on stilts the area under the house could not be seen from the main living area above. The owners had installed a color infrared security camera so residents could see the bottom area on a monitor upstairs in one of the bedrooms.

The family settled in along with the dog and they spent the next day at the beach.

Around 12:00 on the second night the guest dog had barked as if it had seen something. The family viewed the monitor but saw nothing. While it was dark, the night vision camera and low lights on the street illuminated the area under the house enough so that it was a clear picture on the monitor. The picture was of course like those seen in military footage in that it was night vision images. No having seen anything unusual, the family assumed the dog was barking because she was in a new location and was frightened to be alone. The dog usually slept in the house.

On the third night the dog once again started barking at about the same time (12:00) and once again nothing was seen on the security monitor. By now the family had decided to ignore the barking that stopped quickly and was not a problem to them or the other occupants of the nearby homes.

On the fourth night the dog did not bark and the family felt confident their dog had adjusted to the new environment and they could finally sleep through the night without being awakened.

During that night the mother went to the kitchen area to get some cold water, as she was thirsty. As the mother passed the bedroom with the camera she noticed something moving under the house. The mother woke her husband and told him to come look at the monitor.

On the monitor they saw a glowing round disk shaped object flying approximately five feet above the concrete area in the fenced in dog pen. The object was flying around the area as if scanning just as a military aircraft would fly on a mission.

On the monitor the couple could clearly see that the dog was totally asleep and did not realize something was under the house. The husband and wife were surprised because the dog had always been very sensitive to anything unusual and a great watchdog.

As they watched the disk rotated from parallel to the floor to a vertical position making the disk perpendicular to the floor (ninety degree angle) giving the object the appearance of moving and flying sideways. In the words of the couple the object looked like a glowing dinner plate resting on its edge but flying around the area.

As they watched the disk resumed the flat position and stayed in the same area as if hovering. Soon two, three, and then four more disks joined the first disk. After the disks hovered together they went in opposite directions exploring the area.

All of the disks seemed interested in the dog and all hovered just over the dog's head, body, and flew in and out of the doghouse.

Since the dog had not awakened the wife asked the husband to go under the house and see what was going on and see if the dog was all right.

The husband took a small flashlight but did not illuminate the area.

Once the husband was under the house he looked but could not see the glowing disks. The husband checked on the dog but the dog was in a deep sleep as if to be drugged. The dog was finally awakened and determined to be fine but immediately went back to sleep when the husband left the area under the house.

When the husband came back upstairs he informed the wife that the disks had gone. The wife then told her husband that the disks had not gone and told him to look at the monitor. The disks were still flying around the area.

When asked if the disks left and came back the wife told her husband that when he went under the house the disks were flying around him and hovering over him as he moved. The husband said that it not possible disks were flying around him because he saw and felt nothing.

The couple continued to view the security monitor. At one point one of the disks came directly up to the camera lens and hovered sideways in front of it. The couple had a strange feeling the disk knew they were watching and wanted them to know.

After a few more minutes the disks flew away together at a high rate of speed.

The next day an examination of the area showed no evidence of the event and the disks were not seen again for the rest of the vacation week by any family member watching the monitor upstairs.

I did view the video of the incident a few months later but unfortunately the video hard driver storage was automatically erased not long after I saw the video due to storage space being full. The auto erase function triggered before the security access code for the hard drive could be obtained from the owner of the house and before I was able to make a copy of the video.

As I had never seen such a thing before I researched the description of what I had seen. I found references to objects called "critters". Many witnesses have seen critters and according to many sources they can only be seen using night vision or infrared devices or cameras. The descriptions in most cases were exactly what the family saw and what I also saw.

The theories on critters range from UFO's to life forms that live in another dimension or forms that are around us all the time but invisible to our limited vision.

I must admit the theories on critters seem a bit hard to believe but I have seen them myself. There is absolutely no doubt the images I saw were real or that they were not anything one would expect to encounter during the day or night.

The movements of the critters suggested purpose and a high intelligence at least equal to that of a dog or even a person.

Fortunately critters are harmless and seem to have no effect on their surroundings.

The beings or objects called critters do exist, of that I have no doubt.

1700's Man, Best Friends Forever, Flower Girl

This investigation is actually three different stories that are tied together because they occur in the same house.

The house in this story was built in the early 1950's and has stood the test of time over the many years.

Unfortunately the house has been rented several times before being purchased again in the 1980's and has not been treated as well as other houses of its age.

The current owners purchased the house from the parents of a woman who had just graduated from college.

The house was a graduation present for the daughter however the daughter was diagnosed with a fatal disease and never moved into the house. Rather than keep the house the parents decided they would sell it to the current owners.

The current owners repaired the wear and tear on the house and have updated the house several times over the years. Currently the house is very well maintained and has the appearance of a newer house with a landscaped yard. The neighborhood the house is located in has become very exclusive and desirable due to its location next to very expensive homes and locations.

From the start the man and wife would hear noises in the house. The noises would be like someone walking around or moving items in the house.

On several occasions the noise was very loud. The sounds would come and go at random with no real pattern or specific time.

The couple described the noises as sounding like someone else was living in the house with them.

Guests who spent the night in the house also commented on the noises they heard and what might be the cause. The owners would tell the guests it was the house settling and laugh.

There were no incidents other than the noises until the sightings began.

On a regular basis one or both of the occupants of the house saw shadows of someone moving around the house. Not wanting to keep it secret any more the couple discussed seeing shadows with each other and were both relieved that it was not their imagination.

The family dog was a border collie. Border collies are known for their intelligence and ability to reason similar to people.

On multiple occasions the dog would bark and look toward a point in the middle bedroom as if it was seeing something. When the owners checked there was nothing in the room. The sighting by the dog always seemed to occur in the same room with the dog going to the same location in the room each time.

The large living area is separated from the main house by a stained glass door that is kept closed most of the time.

The homeowners began to notice shadows of footsteps under the crack of the glass door and a shadow moving across the stained glass from what appeared to be the hallway.

I was asked to check into the situation by a friend of the couple. I do not chase ghosts or try to exorcise them; my interest is in investigating causes of events that cannot be explained. What someone does after I advise them of my opinion is entirely up to them.

With security being a priority there are many devices to help protect businesses. One device used to protect industrial secrets is called an EMF. EMF stands for Electro Magnetic Field meter and they are used to detect magnetic fields from power lines, equipment, transmitters and other devices that emit EM that might harm people in a work environment. Another use for EMF meters is to detect hidden wireless transmitters or as most people call them "bug detectors".

It has been demonstrated for years that an EMF meter has the ability in some circumstances to detect spirits or ghosts, as they have been known to emit a type of detectable magnetic field.

As a research scientist and investigator, I have in my possession several EMF meters. I decided to take two different types to the couple's house and do a sweep to see if I could pick up any readings from whatever was roaming the house.

At this time the couple had not told me of any particular spot where the events seemed to be the strongest. I went through the entire house with both detectors (one being extremely sensitive) and found nothing except in one area. In the middle bedroom in one particular spot there was an extremely strong reading to the point of being almost off the scales using the sensitive meter and a strong reading using the less sensitive meter.

The couple was very excited as they pointed out that was the exact spot where the dog would always go and bark as if something was there.

I swept to the ceiling and to the floor. The readings were approximately 6 feet high and 18 inches wide and that by chance is the size of an average man standing. I checked under the house in the spot and found no reading. I checked in the attic over the spot and had no reading also. The readings were as if the meter was detecting a person standing in that particular spot.

There was nothing visible or unusual about the area such as temperature difference or other indications of an anomaly.

I asked the couple if I could come back again at a later time and they agreed.

Two months later the couple was going on a business trip and asked if I would like to examine the house while they were gone. The dog was left with a friend while they were gone.

I spent several hours sweeping the house and the only reading on the EMF meter was still in the same spot.

As I was walking in the hallway past the middle bedroom I noticed movement in that room. I turned quickly and saw a definite figure of a man. The man appeared for only a split second but was visible long enough for me to see him entirely.

What I saw was an older man appearing to be in his late 60's or early 70's. The man was dressed in colonial clothes as were worn during the late 1700's. The clothes were dark with a light colored shirt. The man had dark and gray hair and it was long as was the fashion of that time. The man never looked in my direction but appeared to be looking toward the rear of the house from the bedroom.

When the couple returned I met with them and told them what I had found.

We were all surprised in that everyone was sure the entity in the house was the young lady who had died and was not able to move into the house.

A few years went by with the same events in the house repeating them with no particular pattern.

The dog always slept in the large room under the wide screen television and the dog's bed was there for many years prior to the pet passing away. While the dog was sleeping, she would turn over during the night by throwing herself over and making a loud thump on the floor. As all dogs do, she would bark if she needed to go outside during the night. As a resident of a large city the dog was required to have license and rabies vaccination tags on her collar. As the dog was used to wearing the collar it was never removed except for bathing her.

The dog had become very old and sickly and had to be put to sleep. The remains of the pet were cremated and are kept in a wooden box in the same large room where the dog spent most of its life.

The couple talked often about the dog and how much she was missed after fifteen years as part of the family.

On several occasions the couple was awakened by the sound of the now deceased dog hitting the floor as she turned over during the night and the jingle of her collar. On many occasions the couple was awakened by the sound of the dog barking in their house.

While many would say it was simply caused by grief over the loss of their beloved pet, it would not explain how two people would be awakened at the same time during the night by the barking of a dog that was not in that room.

Later the couple would hear sounds of a collar and tags on a dog and the sounds of a dog's paws running across the floor.

The couple was not frightened; they were pleased to know that dogs may in fact live past their death as with people.

The homeowners requested that I visit the home again and scan the living area for EMF traces of their pet. I scanned the area but did not find any readings. Since I was in the house I scanned the middle bedroom again and found the exact same readings in the same area as before. I remember joking to the couple that they were running a bed and breakfast for spirits but they were not amused.

The "colonial man" continued to roam the house until three weeks after my return investigation regarding the dog.

The couple was awakened at night by the sounds of a dog attacking and the footsteps of a large man. The husband armed himself and investigated only to find nothing. The couple stated to me it was the strangest event of their life even though the house seemed to be full of strange events.

Since the night of the dog attack sounds the "colonial man" has not been seen or heard from.

The theory is that once the dog was deceased the spirit of the dog was able to do its duty and attack the "colonial man" and drive him from the house.

I took readings in the middle bedroom where "colonial man was and the readings are no longer there.

Once the events involving the "colonial man" stopped they were replaced by a new event.

The owners began to notice the smell of gardenias in the house. The odor would move around the house and had no pattern to when they appeared or went away.

Both the noises from the deceased dog and the smell of flowers continue in the house today.

It appears "colonial man" was keeping another spirit from inhabiting the house. When the deceased dog was in the same world as "colonial man" and was able to chase off the "colonial man", the entity that was kept away was able to occupy the house. Whether the entity with the odor of flowers is from the 1700's or is related to the situation of the young woman who died before moving into the house is not known.

The family living in the house has no plans to move and say the house has a feeling of "love" that can be sensed.

Next Patient Please!

Jordan did not like being employed by others so he felt it was time for him to start a new business.

At first the business did not do very well. After changing the business many times the business did begin to make a profit and grow.

Each time the business and profits would grow Jordan would move to a larger building to accommodate the growth.

In a time period of ten years Jordan moved his growing business five times.

This story is about the next to last move for the business.

Jordan would travel past a particular building in the downtown area of the city where his business was located. For some reason Jordan was attracted to the building and hoped one day it would become vacant and he could rent or purchase the building for his business.

As he had hoped, Jordan saw a For Sale sign in front of the business. After a tour with the real estate agent of the empty building Jordan knew it was perfect for his business.

The building was built in the early 1960's by a doctor as doctor's offices. As was common for that time period the building is brick with a concrete floor and reinforced concrete roof. The windows in the building have bars on them put there by the doctor to prevent theft of drugs and other similar items.

Jordan had heard of the doctor but had never met him. The doctor had a reputation for being kind and friendly and was well liked in the community. The doctor had passed away a few years earlier and his family rented the building rather than sell it because the doctor had the building built for himself and practiced medicine in the building for over forty years.

After Jordan purchased the building and moved the business into that location, Jordan did some research on the previous tenants.

The previous tenants were in the building for two years. According to the information Jordan found the previous tenants did not have a license to operate a business in that building. There was also an incident where a stranger came into the building and severely beat a secretary. The tenants moved to another location.

Jordan and his employees enjoyed working in the building. The building had a good feeling when you entered and was in a good location for the business and the employees who all lived fairly close to the building location.

As all business owners do, Jordan would frequently work after hours at the building. On many occasions Jordan would hear noises in the building. At times it would sound like someone was in the building opening and closing desk drawers and moving around items. Employees complained on a regular basis that items in their office had been moved and their desks had been opened.

The noises did not bother Jordan except the noise being so loud as to require Jordan to go through the building to see if everything was in order. On at least two occasions the noise sounded like a car had struck the building. When Jordan went outside to check on those noises everything was as it should have been and there was no car or evidence of an accident.

It seemed the building was restless but since it was not disrupting Jordan did not mind.

After being in the building for six years Jordan decided it was once again time for the business to move only this time it was due to the slowing economy and the need for smaller space.

The building was put up for sale through the same real estate company and due to the great location the building was sold in just a few weeks.

The new buyers wanted to tour the building with Jordan so they could see the inside of the building and be advised of issues with the building. There were only two keys to the building and Jordan had both sets.

Jordan met the new buyer and his top employee at the back door of the building to give them a tour. From the first time he met the top employee of the new buyer Jordan knew that person was very unfriendly and pushy. During the tour of the inside of the building the top employee became so abusive and pushy that Jordan almost got into a physical altercation with him only to be stopped by the new buyer who swiftly took his employee away and apologized for his actions.

Since new occupants were moving into the building Jordan had to move his company from the building to the new location. One of the demands of the contract was that the building would be ready to occupy only two weeks from the closing, which was itself scheduled in just one more week. Jordan did not like the accelerated time frame for moving out but that was part of the contract and he had no choice.

As a medical facility the doors were steel and had special security locks. The locks were so secure that no locksmith could reproduce the keys. The locks themselves were of special steel and almost indestructible. Jordan knew about the security of the doors and locks because one of his employees had locked themselves out of the building leaving both sets of keys in the office. The employee was the only person working that weekend and called a locksmith to get back into the building. When the locksmith arrived they informed the employee that nothing could be done to get inside the building short of smashing through the doors with something. The locksmith left telling the employee he wished her good luck. The employee remembered she had left a small front window open and that window did not have bars. A hardware business less than a block away was open and allowed the employee to borrow a ladder. The employee used the ladder to get to the window and she was small enough to crawl inside. The problem was solved but everyone knew from that point on that only a key would get anyone inside the building.

Jordan advised the real estate agent that he would not hand over any keys to anyone until the building was empty since there was valuable inventory and confidential material inside the building.

Jordan arrived at the building late on a Saturday to continue boxing and moving items. According the Jordan the "feel" of the building had changed from one of happiness to anger. Jordan assumed it was his imagination and was due to the business having to downsize due to the economy.

That Saturday night the usual noises were very violent and louder than usual. Jordan had to check the building several times in an hour due to the noises sounding like someone was breaking into the building. Jordan finally decided to leave the building and come back later.

When Jordan and some employees came to work on Monday it looked as if vandals had been busy in the building. Boxes and items were thrown around, items were out of place, all of the lights were off, and all of the doors that were normally left open were shut. Jordan and the employees searched the entire building but found no evidence that anyone had broken into the building. Jordan knew what had happened but did not want to share it with his employees.

For some reason Jordan had the feeling that the doctor was angry and did not like the new tenants. What was a peaceful occupant of the building had become very violent.

Since Jordan and the real estate agents were friends Jordan shared his stories and concern that the doctor did not like the new owners. The real estate agent responded by saying that he still talked to the doctor's family and the doctor's wife had told him for the first time she was having nightmares about her late husband and the building. In the nightmares her husband was very angry that the building had been sold to new owners. The wife actually considered asking Jordan to not sell the building to the new owners but realized it was not her business and people might think she was disturbed.

The move was completed later that same week and the new owners immediately moved into the building. Once the new owners moved in serious incidents started.

Within one week all of the bushes and landscaping that had been in front of the building for decades turned brown and died.

The lights in the building began to malfunction coming off and on for no reason.

Rust starting coming from the water pipes but the plumber could find no reason.

Cracks appeared in the brick walls and fire ants were building mounds inside the building and attacking people despite numerous calls to exterminators who could never find out how the ants were getting into the building, through the concrete floors, an building huge dirt mounds on the carpet.

There were several auto accidents in front of the building on what was normally an accident free street.

A mentally ill man was found standing at the front door with a large butcher knife in his hand and was arrested after the police were called. The man claimed a voice told him to kill everyone inside.

Several of the employees of the new owner quit their jobs for no reason.

Each morning the building appeared as if someone had broken into the building and vandalized it.

A family member of the building's owner drove her automobile off a tall bridge and was killed. No reason was ever found as to why the accident happened or why the bridge railing failed.

Apparently the new owner suspected the truly bad attitude of his top employee was offending the doctor. The top employee was abruptly terminated but the problems continued.

The only thing the new owner could do was to move out of the building. The owner rented a building two blocks down from the doctor's building and put the doctor's building on the market for sale. The owner's business began to decline.

A developer had planned to purchase the building and construct expensive condominiums on the property but the developer contracted an illness that could not be diagnosed and died. The development stage completed by the developer next to the doctor's building for some reason will not sell and they are in disrepair.

The doctor's building is still vacant and has not been sold after almost two years. Despite repeated attempts by experts to solve the problem any plants used for landscaping die once planted around the building. The building has leaks in the roof that cannot be stopped in spite of all efforts and a new roof.

Nearby businesses have quit calling the police due to the lights in the empty building coming on and going off from room to room as if someone is in the building. In general everyone in the area avoids the building to the point that many will not walk near the building but will cross the street and then cross back later to avoid walking in front of the building.

Jordan always takes another route when he is in the area so he will not have to look at the building.

I have not spoken to Jordan, the doctor's wife, or the new owner of the building since my investigation. The real estate agent has closed his business.

Screaming night woman

On the edge of the city limits there is an upper middle class neighborhood. The area was developed in the 1970's and continues to this day due to the large amount of land.

The neighborhood is off a main road and the streets are lined with large houses and streetlights.

Starting in the late 1970's neighbors have been awakened in the middle of the night by the sound a woman running down the middle of the oldest street screaming in horror. The sound is very distinctive and repeats itself several times before the screams stop.

The first time the screams were heard was in 1977. It was early fall and the temperature was mild. Homes in the neighborhood had their windows open for the cool fresh air especially the second floors of the two story homes.

Neighbors rushed to their porches and turned on lights to see if they could determine who was in trouble and if they could help. The police were also called in case they were needed.

By the time the police arrived at the street all was quiet except for the neighbors gathering and discussing what had happened. A search of the area and checks on each house on that street revealed nothing out of the ordinary. The police filed a report and left the scene.

Neighbors assumed someone on the street had some type of mental problem and had not taken medication as needed. That would explain why no one knew who screamed. The family members of the mentally ill person did not want neighbors to find out about their loved one.

The Screams are not heard on a regular basis and are not heard except in the early fall.

The street where the screams originate is the first street built by the original developer in the early 1960's and except for a few older homes on what was previously farm land, the houses on that particular street were the first built.

Some current neighbors think the screams come from a house with a special needs child. The parents of the child had built a special building attached to their home so they could care for the child.

Other neighbors think the screams are from a television or someone trying to scare the neighbors with a practical joke.

If it was someone who lived on the street that would not explain why the screams still occur at times. Due to the type of neighborhood it is, most people live there for the rest of their lives and it has been that way since the neighborhood was first built. All of the original residents who were on the street at the time the screams were first heard have either moved, been placed in assisted living homes, or died.

The screams cannot be from the special needs child because the child was not even born when the screams first happened and she passed away almost ten years ago.

As for television or radio being the cause that could not explain the same screams over a period of over thirty years. The practical joke theory is also not realistic in that no one plays the same practical joke for decades and that person would also be gone from the street by now.

There is a believable explanation for the woman's screams still heard in the middle of the night.

When the development was first started ten homes were built. The first homes sold quickly during 1963 and 1965 and construction was started on phase two on that street.

According to one of the first residents who now lives in a small apartment one of the first houses was purchased by a man and his wife. The man had recently retired from the military after serving in WW2, Korea, and Vietnam.

The couple lived in the house alone as their two children were over eighteen years old and had left home earlier for lives of their own.

In the middle of the night in early fall of 1967 the woman awoke sensing something was not right. The woman turned to her husband and could not awaken him. The wife found her husband had died in bed in his sleep. In a panic the woman went screaming down the middle of the street in her nightgown. The few neighbors living on the street heard her screams and came to assist. The neighbors tried to calm the woman and called an ambulance and police. There was nothing that could be done for her husband, as he was deceased. Apparently her husband had died of a heart attack most likely from heavy smoking.

The widow continued to live in the same house. Due to her grief over the loss of her husband the woman began to drink heavily and was taking sleeping pills.

In the early fall of 1977 on the same date she had found her husband dead, the widow committed suicide with an overdose of sleeping pills and alcohol.

The neighbors did not know the woman had died until they noticed her newspapers were piled up in the front of her driveway and that they had not seen the woman for at least a week. The woman's car was still in the carport and had not moved in over a week. The police were notified and when they arrived and looked through a window they could see the woman lying in bed unresponsive to their attempts to get her attention.

The house was later sold by the children and since has had two residents neither of which said anything unusual happened in the house to them.

The last resident has now sold the house and a new family has moved into the house.

One of the newer neighbors on the street was asked if he had ever heard anything usual at night. The neighbor stated that he had not heard anything unusual and explained that he never opened his windows at night so he could not hear anything. He said the house had new double paned windows installed a few years ago so the house was almost sound proof.

The neighbor was asked if the reason he never opened his window at night was due to potential crime. The neighbor replied that the neighborhood was very safe with no major crimes of which he was aware.

The neighbor then stated that the reason he did not leave his windows open was that some woman living on the street would scream horribly in the road in the middle of the night. He looked out the window but could never see who it was. The neighbor said the scream was terrifying and he did not want to hear that scream again.

Bat Magnet

Native Americans believe that everyone has a spirit guide to guide them through life. Those guides are animals and are reported to come to the people in dreams or in visions.

Bill is not Native American as far as he knows. Bill is just an average guy who is married and has a family. Nothing strange had ever happened to Bill until the events that started in December of 2005.

In early spring of 2005 Bill's father passed away after a brief illness. Bill had spent most of the year taking care of his mother and helping her adjust to the loss of her husband. Bill's parents had been happily married for almost fifty years.

By December of 2005 Bill was finally able to relax a little as most of the paperwork and other issues related to the death of his father were almost complete. Since it was almost Christmas, Bill was looking forward to relaxing with time off from work and enjoying the holidays. Bill knew it would be tough to celebrate Christmas without his father for the first time in his life but Bill was optimistic. Bill and the family had also planned a trip to the mountains for which they would leave on Christmas day after opening presents.

Christmas December of 2005 was cold for the area where Bill lived. Bill and his wife went to bed just before midnight and both quickly fell asleep.

An hour after he fell asleep Bill was awakened by something but he didn't really know what. Bill heard a noise as if the ceiling fan over the bed was running. Bill looked up in the mostly dark room and saw the blades of the ceiling fan rotating and could hear the noise from the blades as the air rushed around them. Bill could not understand how the fan could be running since it was on a separate switch from the light and he was sure he did not turn the fan on.

As Bill became fully awake he looked up again. A stunned Bill realized the fan was not running, what he was seeing was a huge bat flying around the fan blades in a circular pattern. The bat was so large Bill could hear the wind from the bat's wings and could feel the air since it was right over the bed.

Bill quietly awakened his wife and told her not to panic but there was a huge bat flying over the bed. Bill's wife screamed and put her head under the covers hiding in the sheets.

Bill slowly got out of bed and went to the bedroom doorway. When he reached the doorway Bill turned on the lights and saw a giant bat flying around over and over again in a circle just above the bed.

Bill was not one to kill animals however he was worried for his wife since the bat was so close to her. Bill knew if he turned on the ceiling fan the bat could get hit and fall into the bed and possibly bite his wife. As an avid hunter and outdoorsman, Bill knew that bats could carry rabies.

Bill told his wife to stay under the covers while he shut the doors to the other bedrooms so the bat could not enter those rooms. After quickly closing the other bedroom doors, bill grabbed a broom and tried to hit the bat. The bat flew directly at Bill and went right past his face. The bat flew into the great room area. Bill looked for the bat for hours but could not find it. Since the next day the family was leaving for their much-needed vacation in the mountains, Bill sealed off the great room with sheets over the entrance that did not have a door and closed the other doors. Bill thought he would deal with the situation in the morning.

Bill knew from education that the Center for Disease Control states that anyone who wakes and finds a bat in an occupied tent or home should get a rabies vaccination. Bats have tiny teeth and people who have been bitten or infected with rabies may not even know they have been bitten or have any visible marks.

Bill knew the rest of the family was safe. The matter was discussed between Bill and his wife and they both decided they would not get rabies shots even though rabies is fatal once it infects a person and there is a strict time period in which the vaccination must be given.

The next day Bill could find no trace of the large bat. Using duct tape and sheets, Bill sealed off the great room fully so they could go on the trip to the mountains. Bill was sure the bat had not escaped and must still be trapped in the large room possibly hiding in or behind furniture, pictures, drapes, or many other places a bat could hide. Bill read that if you leave a window open a trapped bat would fly away the following night. Bill left a window open in the great room as an exit for the bat.

The trip went well and when Bill returned he could find no trace of the bat. Bill closed the window.

The house also had a recreation type room. Bill and his family used that room as a game room and a playroom. The night before New Year's Eve Bill went into the game room to enjoy some music. It was about 10:00 P.M. when Bill went down the four steps into the game room and noticed a bat flying around the ceiling fan in that room. Bill backed out slowly and looked through the doorway at the bat. The bat Bill saw this time was much smaller and was definitely not the same bat. Bill used a broom to swat the bat after the bat clung to the wall tired from flying around for so long. Bill took the broom and used it to haul the bat outside where Bill placed in on the back patio. The next morning there was no trace of the bat.

Bill inspected the house and found no entrance for bats. The only possible entrance was where the ceiling fan wiring went into the ceiling and Bill used putty to fill that hole inside the cap to make sure nothing could get into the house again.

Bill had no trouble with bats again in the house. While moving the furniture for his wife to clean the carpet, Bill found the dead body of the large bat. The body was perfectly preserved apparently due to the fact the window had been open and the room was very cold the week they were gone to the mountains. Bill was of the impression the bat had clung to the bottom of the sofa which is why Bill did not find it on Christmas Eve when he searched the room.

Bill discussed the bat situation with a wildlife expert who told him bats tried to get into warm areas during cold nights so that was why the bats were trying to get into the house.

In February of 2006 Bill and his wife were visiting his mother and were spending the night at her house. Bill's wife had gone to bed but Bill was not tired. Bill went into the den to watch television and relax. After an hour Bill looked up and saw a large bat circling around the ceiling fan in his mother's den and over the couch where Bill was lying as he watched television.

Bill grabbed broom from the kitchen pantry and swatted the bat knocking the bat unconscious. Once again Bill used the broom to carry the bat outdoors where he put it on his mother's patio. The next day the bat was gone.

The only way the bat could have entered the home was when the door was open the night before and Bill and his wife were bringing their overnight bags into the house.

Bill thought this bat situation was very strange and mentioned to his wife that he believed he was being followed by bats.

The next summer Bill and his wife went fishing at a nearby lake. It was getting dark so the couple packed up and prepared to head home. Once they were inside their car Bill noticed there were several bats flying around his side window. Bill quickly closed the window so no bats could enter the car.

On another occasion Bill was walking his dog when several bats flying in the area came to where Bill was standing and circled over his head. The dog was frightened and ran for home. Bill slowly walked back to this house with the bats circling and following him close to this head. Bill made no actions to threaten the bats but quickly went inside of his house.

According to many Native American tribes, the bat is the guardian of the night. In some other spiritual beliefs bats represent a Karma that has not been returned to a person for something they have done.

Some have suggested as a scientific explanation that when Bill breathes he emits a high-pitched sound that attracts the bats while others suggest there is a more mysterious reason why the bats are attracted to Bill.

To this day Bill is not able to go outside at dusk or when bats are in flight, as the flying bats will come to his location and circle over his head. Bill cannot visit any live bat exhibit as the bats react and gather at the screen or glass closest to his location. At one zoo with live bats, Bill was asked to leave as the reaction to him by the bats was scaring the children.

Bill has spent many hours examining the area around the entire outside of his house in order to screen off or block any possible entrance for bats. Bill also makes sure he never leaves any outside doors open for more than a few seconds after dark and carefully watches to make sure that a bat has not flown past him as he enters or exits the house at dusk or after dark.

Bill and his wife are fine with regard to their health and fortunately they never contracted rabies from the incidents.

Radio stations

From the time he was a young man, Robert would write songs in his sleep. At least once a week he would wake up with a song in head complete with words and lyrics. Robert assumed he was a gifted songwriter. The Problem is Robert could not write the songs he simply dreamed them and woke up with the tunes in his head but could not write them down before he forgot.

It wasn't until a few years ago that Robert realized why he could write songs in his sleep. Robert has an unusual talent or gift as some might call it, Robert can hear radio stations in his head without a radio.

It was by accident that Robert discovered his strange gift. While riding with a friend in a car Robert started singing a song. Robert's friend did not want to hear Robert sing so he turned on the car radio. The song playing on the radio was the same song Robert was singing. Both Robert and his friend thought it was just a coincidence.

On another day, Robert was singing a song while preparing his breakfast before classes and when Robert turned on the radio the same song was playing. Robert decided it was just another coincidence.

Robert was at a party with his fellow college students later in the semester when Robert's friend jokingly told other people that Robert was psychic and could hear songs on the radio while they were playing. The crowd laughed and then demanded a demonstration. Robert at first refused but since everyone was expecting a good laugh at his expense, Robert agreed.

The crowd asked Robert what song was playing on the Radio. After a few seconds Robert laughed and named a song. Quickly someone ran to the stereo system and turned it to FM radio, the song Robert named was playing. Most laughed at the coincidence others seemed to look puzzled. The partygoers asked Robert to do it again but Robert thought he should quit while he was ahead and said he was too tired to try it again.

The event got Robert's attention even though he kept his interest to himself. A few weeks later Robert woke up with a tune in his head. Robert immediately turned the radio on and the same song was playing. The song on the radio was a new song and was being played for the first time on that station according to he DJ after the song ended. Robert was now very interested in his situation.

Robert would repeat an experiment for many months. In the experiment Robert would sing any song that came to his mind and then turn on the radio to see if it was playing. According to Robert over eighty percent of the time he was able to predict what song was playing on the radio.

As if Robert's situation was not strange enough, the issue of how does he know what song is playing on what station came up in a conversation. Robert stated he hears songs as they are being played on the station that is set on the radio nearest him. It does not matter if the radio is a digital tuner or an old style tuner.

Several times Robert conducted an experiment related to his theory about nearby radios determining what song he would hear. When Robert would hear a song in his head he would turn on the nearest radio only to hear the same song playing on the radio that was playing in his head. Robert would quickly turn to another station to make sure the same song was not playing and there was always a different song on the other station.

Robert never listens to country music but for a month he set all of his radios on the same country station. For that same month all of the songs Robert heard in his head were country songs.

Robert then set all of the radio stations in his apartment to classic 60's music and for that time period only heard 60's classic music that was playing on the radio at the same time.

It does not matter where the radio is or what type it is, Robert seems to have the ability to hear the radio broadcast without a radio being on. Robert can also hear talking and news broadcasts but for some reason cannot understand the words but he is able to know that a song is not being played at that time on the radio station.

Robert clearly states that there is no use for his unusual talent and that he has no control over it. Most of the time Robert can hear a song but other times he has no idea what is playing on the radio. Robert cannot determine what triggers his ability.

There is more to this story that is even stranger than hearing radios that are not on.

Robert decided he did not want to hear radio in his head without the radio being on so he removed all of the radios from his apartment. The lack of radios did help some in that Robert would hear songs and broadcasts less frequently and they would also start and stop in the middle of the song.

On a summer night, Robert woke up hearing a song in his head only this time it was more clear and had a tinny sound like a small radio instead of the usual deep rich sound he usually heard. In the darkness Robert listened and turned his head. As Robert turned his head he noticed the song got louder and he could hear it more clearly. It was at that time that Robert realized he was hearing the song from the window air conditioner unit that was running in his bedroom.

Robert laid in bed in the dark and could hear the radio station as clearly coming from the air conditioning unit. Robert got used to the situation and it did not really bother him the rest of the summer.

Winter came and Robert once again was free of the radio station broadcasts in his head and the songs. Unfortunately the silence ended when it became cold enough for the central heating unit to run. Each time the central heating fan would come on Robert would once again hear radio broadcasts that would stop when the blower fan shut off. With the furnace fan Robert could even hear talk radio shows but could not make them out just as before with radios.

Robert now realizes that almost any electronic or electrical device can be a conduit for his mind to receive the songs. A simple item such as a microwave running, an electric razor, or even a computer-cooling fan can trigger Robert's ability to know what song is playing on the radio as it plays. The only difference with the devices and a radio is that Robert hears the radio station with the closest transmitter on the device rather than one tuned on a radio.

Robert has tried to convince himself that the radio station is simply transmitting at high power and the wiring in the motors or circuits of the devices is acting as a radio tuner. The problem with Robert's theory is that he has asked his friends many times and none have ever heard the song playing or any radio broadcast when Robert hears them. A friend of Roberts who is studying engineering even brought a high tech sound recorder but did not record any radio sounds even though Robert was hearing them. Robert has now stopped asking his friends about the sounds so they will not think he is losing his mind.

A girlfriend suggested to Robert that he purchase a sound machine to drown out the radio station sounds. Robert did purchase a sound machine that worked for a while until he started hearing songs within the sounds coming from the sound machine.

Robert has learned to live with his unusual talent as he simply leaves a radio playing in his apartment and car at all times. Robert has found a way to defeat what he calls the "curse of the radio man".

The Glow

Abandoned houses in the country seem to be the favorite places for spirits or ghosts to inhabit.

One abandoned house in the country proves that some people never leave home even after death.

Off a major highway approximately twelve miles from a large city sits an old abandoned farmhouse. The house is unpainted wood that has turned dark to the point of being almost black and small bushes have grown into trees over the many years. Most of the windows of the house are broken or missing. The farm fields around the house have long since become overgrown with trees and wild grasses. The tin roof still protects the home interior somewhat but is also failing. The house itself sits on a small foundation of mortar and granite stones. Based on the design of the house and other available information the house is circa early 1900's.

Sam who was the owner built the house. Sam's wife was named Violet. When the house was built electricity was fairly new and not yet available in the country so the house was never wired and Sam used oil lamps to light the house and wood to heat the house in a wood stove located in the middle of the house. Sam's occupation was farmer and he grew several types of produce that he sold at the local farmer's market. Sam would also hunt deer on his property during deer season. The dirt road that leads to Sam's house actually has his name as the name of the road.

The residents of the farmhouse have long since died leaving no children or family members to inherit the house and property.

The county has tried several times to sell the house and property for taxes not paid but to date no one has even bid on the property. There may be a reason why no one wants the house.

Sam was a typical farmer. Sam was average height and a thin build with beard stubble on his face, wrinkles from the hard work in the sun, and thinning white hair. Sam always wore overalls with a white tee shirt underneath. According to those who knew him, you would always see a red railroad style bandanna hanging from Sam's rear pocket.

Violet was said to be a very sweet person who always had a smile on her face. Violet was slightly heavy but had a pretty face. Most of the time Violet would be wearing a large apron as she did her chores around the house and fixed meals.

Violet died first in the early 1940's at the home and then Sam died in the late 1940's in his sleep at the age of ninety-two years old.

Sam and Violet were both buried on the property but the wooden markers have long since been destroyed by the weather and due to the dense growth of trees and plants and time. No one can tell you exactly where he or she is buried.

During the 1950's residents of the area first started noticing strange things in the house at night.

At night on several occasions concerned neighbors noticed what appeared to be a fire inside Sam's house. On each of those occasions the volunteer fire department responded only to find no fire and no evidence of a fire.

The false alarms to Sam's house continued for years into the 60's and actually became a type of joke to the Fire Department. Statements like " it is a good night for a run to Sam's house" were common.

As word spread about the mysterious non-fires, local high school students became interested.

Driving by Sam's house and looking for the strange fire was something for high school seniors to do on the weekends and during the summer. On at least one occasion a student told his friends that he and his date had seen the fire inside Sam's house but it went out before they called the Fire Department.

The curiosity of the local kids for Sam's house grew. By the mid 1960's the students and their dates or friends would hang around the area where they could watch the house turning the area into a type of "lover's lane". Once again on a few occasions students told of seeing fire inside the house.

By now the current high school tough guys decided they would find out what was going on and solve the mystery. The tough guys were graduating seniors and were either being drafted or going to college so they wanted to do something to make them famous.

On June 25, 1966 four graduated high school seniors drove their car to Sam's property and waited in the car for something to happen that night at Sam's house.

At 10:00 P.M., one of the four young men noticed what appeared to be a fire in Sam's house. Quickly and quietly they exited the car to catch whoever was inside the house. As they approached the house they could see the fire was actually a lantern not a fire.

The four young men got close enough to the house to see through the windows and saw an old style oil lantern moving around the house but no one was carrying the lantern. The young men watched as the lantern moved to different rooms as if it were floating in the air. The tough guys realized they were not so tough after all and made a quick retreat to the safety of their car. As they got into the car the driver realized the car would not start. As the other three young men yelled at the driver to get the car started one noticed the lantern was now in front of the house and slowly moving towards them as if someone was holding it in their hand and walking towards the car. The car finally started, as it seems the driver had flooded the engine trying to get out in a hurry.

The young men promised each other they would not tell anyone of their experience at Sam's house since it would make them look like scared children.

The four young men decided never to go back to Sam's house again.

Exactly one week later to the day of the event at Sam's house, at exactly 10:00 P.M. one of the young men noticed from his bedroom window a glow coming from the backyard of his parents' house. When the young man went to the window he saw a thin old man, with white hair, wearing coveralls carrying an oil lantern. The man appeared to be looking up at the young man and appeared angry. The young man grabbed a baseball bat and went downstairs to go into the backyard and chase off the intruder. When the young man got to the backyard there was no one there.

The next day the young man called his friends and each told him the same thing had happened to them at the same time.

They all realized now that the figure and light they saw was Sam's spirit getting revenge for their trespassing on his property a week earlier.

Eventually the four young men did tell family members and friends of what happened and suggested they not go near the house again.

Even now the lights in the house can still be seen at night in Sam's house. The area around Sam's house has not been developed and is still mostly farmland and woods as it has always been. The neighbors have told the story of Sam for many years and still tell new residents or young people to stay away from Sam's house and stay off his property.

The warnings have worked in that the house sits undisturbed by people and will probably be that way until time and the elements totally destroy the house.

The question is whether or not Sam will approve of a new house being built on the property or whether Sam is only there to protect the house he built. Time will tell.

The Thing In The Woods

Many people do not realize that almost 90% of land on the entire planet is totally unexplored by man.

There are many unsolved mysteries involving creatures or unknown beings that have never been solved. The Loch Ness monster, Bigfoot, and others are just a few modern examples.

The Southeast coast of the United States still has many unexplored and undeveloped areas. The areas are protected by Federal Laws relating to wilderness and coastal wetlands. Development in the protected coastal areas is not allowed or in some cases may be allowed but restricted as to the number and type of developments.

One such Island exists off the coast of a Southern state. Because the barrier island in this story is mostly wetlands and marsh area, the majority of the island is undeveloped and unpopulated. Like all coastal areas, development is taking place as people become aware of the beautiful and unspoiled island.

With development will also come many discoveries. The discoveries on the island in this story are Indian artifacts, revolutionary war items, civil war items, and wrecked ships exposed as the ocean removes beach sand during storms or hurricanes.

There is one discovery that has been kept secret among those who have lived on the island prior to inhabitants of the new houses being built.

This island has three main areas. The areas of the island are the beach, the marshes, and the wooded areas.

One development was started ten years ago in what was a combination wooded and old farm area on the island. The development has resulted in a quiet neighborhood of less than twenty homes on large plots of land. Heavy woods, tidal creeks, and marsh wetlands surround the neighborhood development.

Even with the development, the neighborhood and surrounding areas are a haven for wildlife such as herons, bobcats, deer, raccoons, snakes, and most other animals one would associate with Southern woods and coastal areas.

The first inhabitants of the neighborhood were mostly local people who grew up on the island but in recent years larger and more expensive homes were being built as beach property became harder and more expensive to obtain.

For many years there had been rumors of a large animal roaming the woods of the island. The rumors were a type of secret kept between the locals so as not to scare off development and the related jobs and money.

Still not being fully developed, the neighborhood has many opportunities for exploring and exercise by the residents.

One pastime is to walk a dog or ride a bicycle around the paved roads in the neighborhood.

It seemed odd to many that no one seemed to walk at night or allow their children to play at night in that area. The explanation by most was that the mosquitoes and snakes represented too much of a danger at night.

It turns out the mosquitoes and snakes are the least of the worries of the people of this island.

People in the neighborhood and surrounding wooded areas all had dogs. The dogs were used for hunting and to warn the homeowners to strangers since the houses are fairly remote and law enforcement is rarely seen in the area.

As is the case with most rural areas, the people would turn their dogs loose at night to exercise and roam as a way to offer the dogs some recreation.

The problem started when dogs in the neighborhood starting to disappear. These were not small dogs these were large breeds such as Labradors, Pit Bulls, Rotwiellers, and mixed breed German Shepherds. The dogs would leave at night and never return.

At first the owners assumed the dogs were bitten by venomous snakes and died. That theory did not last long as the bodies of the dogs were never found.

Another theory was that a neighbor did not like the dogs roaming and was killing them; the problem with that theory is that every single neighbor had a dog and every single neighbor had a dog disappear. According to residents of the area over a three-year time period, almost fifteen dogs had disappeared without a trace.

Several times residents in the area had noticed a strange sound coming from a wooded area near wetlands. The sound was described as a very low growling sound mixed with a higher pitched howling sound. The sound was like no sound any of the residents had ever heard before and most are avid hunters and outdoors people.

On one occasion at dusk a resident noticed a large furry animal run in the woods. The animal was moving fast and the resident described the animal as tall, covered in fur of some type, and very powerful in build.

As the animal noise became more frequent and sightings became more common the US Wildlife department was contacted. When asked what type of animal might be inhabiting the island the officer said there were no such animals on the island and refused to believe such an animal was roaming the island.

The State Wildlife Department was also contacted by some residents for information and that agency suggested it might be remotely possible the animal was a large American cat of some type such as a panther or mountain lion that might have made its way to the island or escaped as a pet. The State Wildlife officials also mentioned it could be a black bear as they have been known to inhabit barrier islands on the coast.

The residents who saw the creature stated they were absolutely sure it was not a bear or a large cat due to the way the creature moved and that is was running on two feet not four feet.

With the loss of so many dogs, the residents started walking their dogs on leashes or keeping them in a dog pen for protection from whatever was killing them.

In early spring a man was walking his German shepherd at around 8:45 P.M. The dog became excited and started pulling the man towards a wooded area as if to catch something. As the man and his dog entered the wooded area there was a strange animal sound. The man described the sound as a howling growl that started very low and became very loud towards the end of the growl. The usually fearless German shepherd ran from the woods to the road pulling the owner with him.

When the man and his dog approached their home several neighbors were in the area. The neighbors stated they heard a man yelling and wanted to know what was happening. It was explained to the neighbors that no one had yelled that is was in fact some kind of animal. The man told the people where the incident occurred and the shocked neighbors could not believe they could hear an animal that far from their homes especially since they were inside their homes at the time. The dog refused to walk in the area again despite several attempts to walk the dog near those woods during daytime. The man has never walked in that area after dark since the event.

One family in the neighborhood owns a very large Golden Labrador male. The Labrador is kept in a pen for safety but managed to climb out one night. When the family discovered the dog was missing they searched the area. As they searched, the family heard a dog fighting and a strange growling sound then a dog yelp as if injured. The father and son grabbed shotguns and portable searchlights to find their dog. The father and son heard movement in the woods similar to a large man running and headed in that direction. When the two got to the area of the sounds they found their dog. The dog was badly wounded and bleeding. The dog was taken to an emergency veterinarian clinic located 45 miles from the island. The vet was able to save the dog but only after the dog received 185 stitches to his stomach.

The father asked the vet what type of animal would cause such a wound to his dog and the vet replied that he had never seen a wound of that type in his 25 years of practice. The vet ruled out a bear and a large cat due to the type of wound and the marks around the wound itself. The wound was so unusual the vet could not determine if the wound was made by teeth or claws but the vet was sure the wound was not from some type of weapon such as a knife, shovel, or similar manmade items.

A month later a homeowner was awakened by the sound of his dog barking in his pen as if very upset or scared. The homeowner grabbed a loaded 9 MM semi automatic pistol and went to see what was wrong with the dog. As the man was walking towards the dog pen he looked in the direction the dog was barking and saw a large furry animal standing on two feet staring at him. The man thought it might be a person in some sort of costume playing a trick but quickly realized this was not the case as the animal growled and started to move again. The man ordered the animal to stop in case it was a prankster and when it did not he fired 10 shots at the animal. By this time the neighbors had heard the incident and gunshots and had called the sheriff's department.

The deputy arrived 10 minutes later and was joined by several other units within another 10 minutes. The deputies searched the entire neighborhood using the high power lights on their cars but could find nothing. There was no blood trail or evidence that any animal or person had been shot.

There are very few dogs in the neighborhood now and the few dogs that are left are kept in pens at all times.

The children of the neighborhood are not allowed to leave their house near dusk or after dark.

There is evidence that something is in the woods around the neighborhood. On several occasions, neighbors have located objects that were causing bad odors. The objects were deer and all had been torn to pieces with few pieces remaining of the large animals.

If you drive through the area at night you will notice something very strange. Every single house in the neighborhood has multiple floodlights around their homes that light up their entire yard. The outside lights on the homes come on at dark and never go off until the sun rises again. The neighborhood is lit up like a football stadium during a night game. It seems the neighbors believe whatever is in the woods is afraid of light.

Perhaps the island has its own Bigfoot type animal or it could be some type of animal no one has ever seen or heard of before. The deer remains are still found by hunters but closer to the unpopulated marsh areas and other wooded areas.

The question remains, what is the creature and what actions will the creature take when it runs out of places to hide or live?

Revenge Of The Insects.

We have all heard the saying "leave it to a professional" but what happens when professionals do not want the job?

All types of insects live around and under buildings. Many insects we never see other insects we do not want to see.

An older house has many secrets than can be revealed quite by accident.

Four years ago a resident of a home had noticed a strange insect in their living room. The light colored bug looked prehistoric with protrusions on its back and long legs with strange eyes. About the size of a grasshopper, the homeowner called the local county agent to inquire as to what type of insect it was.

Based on the description, the agent told the homeowner it was a camel cricket. Camel crickets are strange crickets that have a hump in their back and live in dark places. The agent informed the homeowner the insect was harmless and peaceful and rarely invaded homes where people live and not to worry about finding the insect in the home.

The homeowner took the advice of the agent and did nothing.

When summer came the homeowner decided to add an exterior light pole with an antique street lantern to the side of the house for security and to add a touch of character to the old style house. The homeowner hired a local electrical contractor to do the wiring under the house.

The contractor and a helper crawled under the house. The house was very low to the ground to the point that in some places the bottom of the floor would touch service workers lying on their backs. Since the light was to be located on the side front corner of the house the wiring was required to be run from a switch inside the house to the farthest point from the crawl space access door.

After a few minutes the contractor and his helper came out from under the house and said they could not wire the light. When asked why they could not wire the light the contractor replied that that foundation was covered with what looked like thousands of horrible looking jumping insects.

The homeowner explained that the insects were camel crickets and harmless but the electrician replied "no thank you".

After the electrician left, the homeowner decided to see what was going on under the house. The homeowner stuck his head through the access door and saw no insects under the house. The flashlight was then shined on the brick foundation walls and revealed what indeed looked like hundreds of scary looking camel crickets. The insects were not moving, they were just sitting on the side of the foundation and their eyes reflected the light as if they were all staring at the light and the homeowner. The homeowner made a quick retreat shutting the door behind him.

The next day the homeowner called a pest control company and explained the problem. The pest company sent out an experienced technician who would spray under the house and the foundation. Not long after he went under the house, the technician came out quickly and was slapping his shirt and trying to take his clothes off. The puzzled homeowner rushed outside to see if the man was injured and asked what happened. The technician replied that he had been "attacked" by hundreds of camel crickets. The crickets had all jumped him at the same time and had crawled up his shirt, up his pants legs, and had jumped into his face. The technician said he had never experienced that before but would be back the next day.

The next day the technician brought a fogging type device. The device was plugged into an electrical outlet and a large house was placed in the doorway. The machine was turned on and produced what looked like a fog of pesticide meant to terminate the pesky insects. After the proper time, the technician left and said everything was in order and there would be no more trouble.

No camel crickets were seen again for a time.

Less than a year later the homeowner was sitting on the living room couch talking with a friend when an insect jumped on the homeowner as if to attack. The startled homeowner realized it was a camel cricket and was surprised to see one in the house especially since the pest control man had treated the crawl space and the insects should have all died.

Two nights later the homeowner was asleep and felt something hit his face, then again, then again. Camel crickets were in the bedroom and seemed to be deliberately going for the homeowner's face. The homeowner smashed the insects with his hand and had no more sleep that night.

The next morning the homeowner went to the bathroom to take a shower. While in the shower the homeowner felt something on his back. The homeowner used his hand to hit whatever it was and it was a camel cricket. Another camel cricket hit the homeowner on his leg. The homeowner looked up and the wall near the ceiling was full of camel crickets that seemed to be staring at him.

This particular house had a sunken den and recreation room. The recreation room was built on concrete and had no access to the crawlspace since it was at the same level as the soil at the bottom of the home's crawl space. The homeowner was enjoying a beer and listening to his favorite CD when he felt someone was watching him. He turned around to see a dozen camel crickets on the top back of the sofa staring at him as if they were planning to attack. The homeowner left the room and shut the door trying to determine how the insects got into the recreation room and he could not figure out any way they could have entered the room.

The homeowner was shaken and tried to figure out why the insects had invaded his living area. He had lived there for years and had no problems with camel crickets or other insects.

After much thought the homeowner decided the camel crickets were angry because he had killed so many of them and the survivors were getting revenge.

The homeowner considered calling the pest control company again as many times a needed to eliminate the camel crickets and keep them away but he feared what might happen the next time if any of the insects survived.

Rather than try to kill the insects the homeowner leaves them undisturbed under his house. When the heating/AC company comes to service the system and change the filter the homeowner tells them there are camel crickets under the house and do not disturb them.

If the homeowner does anything that upsets the camel crickets such as a plumber crushing a camel cricket while working, the homeowner will be visited by a single insect that will find him to let him know they are upset by jumping on him and staring at him. The homeowner will always gently cup the insect in his hands so as not to harm it and return the insect outside.

The regular service workers know not to hurt or disturb the camel crickets and have no trouble while working under the house.

The original man that was attacked by the mass of camel crickets left the pest control company after having nightmares for months according to the owner of the pest control company.

So far the homeowner and camel crickets have existed together peacefully now for quite some time with both staying in their living areas.

Curse man.

Most people think a curse is not real or all in the mind of the person who thinks they have been cursed. In the case of the curse man, the facts point away from coincidence and straight towards reality.

Al does not practice any type of sorcery, witchcraft, or other groups usually connected with curses.

Al is just like any other person. Al has a job, a family, a home, and is well liked by all who know him. Al is a friendly person and likes almost everyone he meets. Al is hardly the type of person you would associate with placing curses on others.

Al's ability to curse people did not appear until later in Al's life. He started to notice bad things would happen to people who tried to harm him. Al never wished harm on people, it just seemed to happen and he had no control over what would happen or when it would happen. All Al knew was that at a point in his life things started happening to those who tried to harm him.

Al tried to convince himself that the incidents to those who tried to harm him were just coincidence but after so many people had things happen to them Al finally realized it was not coincidence.

Al first noticed his ability to put a curse on people shortly after he moved to a new house. The new house was located on a busy highway and the road only had one entrance and exit to the highway.

A neighbor had planted many shrubs on the side of his yard. That house was located on the highway in such a manner that the shrubs made it difficult for Al and others to enter the highway because the shrubs slightly blocked the view of oncoming cars and trucks.

Al and others asked the neighbor to please trim the shrubs as someone might get hurt or killed in an auto accident pulling out into the highway. The neighbor ignored everyone's request to help his neighbors avoid an accident.

The highway department and others were contacted about the problem but refused to cut the shrubs as they were on private property.

At one point the shrubs were so large that it was almost impossible to pull onto the highway without risking an accident.

The shrubs were hanging over the road and Al decided he was going to take action.

Al took a pair of hedge clippers and stood on the side of the road and started trimming the parts of the shrubs that were not on private property. The neighbor saw Al and came out to confront him. When asked what he was doing by the neighbor, Al said he was trimming the shrub that was on public property. The neighbor went back in his house and came back holding a gun. Al was not afraid and continued to trim the shrubs. The neighbor told Al that he could shoot him for what he was doing and Al told him that would be a bad idea.

After a tense standoff the neighbor said that he was about to leave on a vacation trip and did not have time to argue the point and he went back into his house. Al saw no reason to call law enforcement since the neighbor never actually pointed the gun at him.

The next time Al saw the neighbor it was almost a month later. The neighbor was brought to his house in a medical transport vehicle. The neighbor had casts on both legs. Later Al learned the neighbor had gone on a ski trip and had badly broken both legs requiring surgery.

While he was recovering, the neighbor became addicted to painkillers. Due to his addiction, the neighbor lost his job and had to sell the house.

The new neighbor is a quite polite person and keeps the shrubs trimmed so as to not block anyone's vision to the road.

Just a coincidence, Al thought.

While returning home from a trip to Florida and driving on the interstate highway, Al and his family were harassed by a driver in a very expensive high-speed sports car. Even though Al was doing above the speed limit and was in the right hand lane it seemed to make the sports car driver angry. The sports car driver would tailgate Al as if to run him off the road, would pass Al and then cut in front of him and slow down. This situation went on for miles. Al realized the driver was dangerous and could hurt his family so Al exited at the next exit to get away from the driver.

The driver of the sports car followed Al making obscene gestures at him and cursing him.

At one point Al lowered his window and screamed at the driver that he was going to get killed playing with people on the interstate. The driver laughed and drove off with another last obscene gesture to Al.

Al and his family found a fast food restaurant and had a nice lunch.

Al returned to the interstate and less than ten miles down the road he noticed traffic was slowing and the cars were bumper to bumper.

After an hour, Al saw the accident that was the cause of the slowdown. Al immediately recognized the car as the one that had been harassing him previously. The high-speed sports car had run off the road and wrapped around a large tree. The deceased driver was so badly mutilated in the accident that the state troopers had put a blanket across the driver's side so people on the interstate could not see the horrible sight.

Al told himself it was just another coincidence.

Al liked his boss at work. Al's boss and he were friends and had known each other for years.

A young friend of the owner of the business where Al worked had been fired and needed a job. The owner made it impossible for Al's boss to continue working at the business and when Al's boss quit his job the owner immediately hired his friend to replace Al's boss. Al was very upset over the situation.

The new boss also had friends that needed jobs due to the economy and his goal was to make life so miserable for a lot of people that they would quit their jobs so he could replace them with his friends.

Eventually the new boss targeted Al and started making Al's life miserable. Al stayed at work and suffered for two years while looking for another job. Eventually Al simply quit his job as the constant harassment and added duties was hurting his health and his family life.

The new boss had a reputation as a womanizer and hired young women so he could date them and soon the business was full of very attractive young single women.

Less than a year after Al left his job he found out from employees who were still working at his old job site that the new boss had discovered he had prostate cancer. The surgery did not go very well and the new boss was no longer able to perform as a man.

Al told himself it was just another coincidence. After all, Al had never hoped anything bad would happen to any of those people so how could he be responsible in anyway?

A new neighbor moved next to Al. The neighbor had several large dogs and decided to build a dog pen outside of Al's bedroom window. Al asked the neighbor to build the dog pen in another location but the neighbor refused.

The dogs in the pen would bark late at night. Since the neighbor's bedroom was on the other side of the neighbor's house the barking dogs did not disturb the homeowner. Al is not one to call the police on his neighbors so Al installed soundproofing on the inside of his bedroom windows that helped a little.

Al's neighbor owned a small business that was very successful. Not long after the dogs were located near Al's window a competing business moved next to the neighbor's business. In a few short months the neighbor's business had to close. Because of the competition, the neighbor was losing the house to foreclosure, and the neighbor moved in the middle of the night so his car would not be repossessed.

By now Al was realizing the events could not have been a string of coincidences.

Other events that happened to people that had tried to harm Al were:

1. A man tried to cheat Al out of money. A career criminal while taking money to his bank later robbed the man who tried to cheat Al.

2. Al's first "love" was a cheerleader in high school. Since Al was not popular the cheerleader pretended to like him to make fun of him and get laughs from the popular kids at school. Al recently found out the girl contracted a type of Staph infection that destroyed her face.

3. A family member tried to stop Al from buying his father's house after his father died by taking Al to court. The family member has since contracted a very rare disease that will result in that person soon being an invalid.

4. Someone who wanted to hurt Al sued Al's wife. The wife of the attorney that handled the attempted lawsuit was arrested for drug use and the attorney lost his license to practice Law due to actions related to the situation.

5. Al traded his car on a newer used car. Al's car was in perfect condition. The car Al purchased developed expensive mechanical problems not long after he bought it. The dealership that sold Al the car refused to pay for the repairs. Al paid to repair the car he bought and not long afterwards the dealership called to complain that a customer had purchased Al's trade in and the engine and transmission seized before the customer left the dealer's lot. Al explained the trade in was in perfect condition when he had it.

There are many more situations related to people that have tried to harm Al.

The good news is that it seems good things also happen to people who tried to help Al.

Since Al was not a member of any cult or similar group he decided he would do some research into his family history. While searching records at the State records facility Al discovered something that he had never known before. Al's great-great grandfather was a Native American medicine man who married a non Native American woman.

Al is convinced that the power of his great-great grandfather surrounds him or was passed on to him and harms anyone who tries to harm him.

Al began to think about stories his father told him regarding his father's combat tour in Vietnam. On many occasions strange things happened to the enemy trying to kill Al's father. In one situation the enemy was firing mortars at Al's father's position and a mortar shell blew up in the mortar killing the enemy and destroying their position. In another event Al's father was caught in an ambush and when approached by the enemy they fired their weapons at Al's father only to have both weapons misfire and give Al's father and his platoon time to kill the enemy. Al's father had many similar unexplained stories he told Al over the years.

Until this case I was totally convinced there was no such thing as a curse. After investigating and verifying the events I am now convinced that there may be some power that can get revenge on people who try to harm particular people. Some people call it bad Karma, a curse, or bad Manitou.

Whatever it is called it is certainly best not to try and harm anyone because it might be Al and Al has no control over what might happen to you.

He Hears The Earth Breathe.

What many call a gift others call a nightmare.

Billy has the unique ability to hear the earth as a living thing.

As strange as it sounds Billy can hear the sounds the earth makes as it maintains us and every other living thing on the planet.

The sound Billy hears is described by him as a low rumbling sound similar to a large factory machine running constantly but coming from inside the earth.

At first Billy thought the sound was coming from cars and trucks on a highway with the sounds being conducted so he could hear them miles away. Billy abandoned the highway theory since he hears the sounds when he camps in woods almost a hundred miles from a busy highway.

Another of Billy's theories was that the sound is coming from power plants producing electricity however he still hears the sound even when he is not near a power plant or any other large noise making location.

Billy hears the rumbling sound coming from beneath his location constantly and describes the sound as a type of gigantic heartbeat that rumbles as if it is alive.

Billy had tried to avoid the sound to determine if it really is coming from inside the earth itself.

Billy arranged to have his hearing tested inside a soundproof room at a medical facility located on the second floor of an eight-story building.

The room itself is used to test hearing and is totally soundproof with a special door, walls, and soundproof glass.

Inside the room Billy could hear nothing except the rumbling sound he hears at all times while awake when there is no other noise to distract him. The audiologist used sensitive equipment on a maximum setting to try and pick up the sounds but was unable to detect any sound coming from inside the booth.

Medical examinations and tests have proved that Billy has no medical problem or ear condition that would cause the type of sounds he hears.

Billy has even chartered a boat to take him miles from shore and drift to see if he can hear the sounds. While the sounds at sea are more muffled, Billy can still hear the sounds. The difference in volume at sea has convinced Billy the ocean is insulating the sound slightly which proves the sound is coming from inside the earth and that there is no other possible source.

On one occasion the sound Billy heard was much louder than normal at the same location. It was the next day from that loud sound that a volcano erupted. Billy has also heard louder sounds from the earth prior to an earthquake but is unable to predict a coming quake. During the Iraq "Shock and Awe" and the North Korean nuclear tests, Billy noticed the sounds from the earth were slightly different and described the earth as being upset or sad.

Billy's situation and the type of sound he hears has convinced him without a doubt that the earth is in fact a living being and that the earth breathes in some way and has a heartbeat just like all other creatures on the planet.

Billy also knows he has an early warning system. When Billy stops hearing the earth living or the sounds change drastically Billy will know in advance that everything on the planet is about to die.

So far Billy has said the earth sounds the same as it has most of his life and at this time he is not worried about the end of the world or all life on the planet.

The UFO Incident.

This incident started in another state and eventually became directly connected with me.

As a digital production expert and as an image enhancement consultant I have the experience, software, computers, and devices to enhance photographs and images. I was able to utilize my resources to help in a determination of the source of images seen over Arkansas.

In early 2007 an ex F-16 fighter pilot noticed lights in the sky that he found to be unusual. The pilot was able to photograph those lights using a digital camera resulting in good quality color images.

After reading the story I decided I would like to use my equipment and skills to enhance the images to see if I could determine if they were in fact UFOs or some other objects. Since the images were posted on multiple news sites, I downloaded the original images.

Using multiple pieces of equipment and imaging software I was able to greatly enhance the UFO images to determine if they were in fact UFO's or simply mistaken objects.

After several enhancements using many software programs the original images and resolution were magnified many times to the point of becoming very clear.

There were two specific images that were the most clear. Upon enhancement one globe contained what appeared to be a pilot sitting behind a control console. The other globe had a similar object inside but that pilot was facing the camera and appeared to either have a triangle shaped head with slanted eyes or the pilot was wearing a type of flight helmet.

I emailed the enhanced photographs to several web and news outlets so they could share the images with others. Soon the enhanced and original images of the UFO's were around the world and can still be located on the web by searching "Mark Kirby UFO".

Not long after the images were provided I was contacted by George Noory for his syndicated radio show called "Coast to Coast" which is also broadcast using his web site. "Coast to Coast" specializes in unusual events and strange occurrences. Mr. Noory invited me to be a guest on his show and I gladly accepted.

I must admit I was surprised at the number of people that listen to or access "Coast to Coast" in that I was constantly being contacted by friends and people I know who told me they heard me on Noory's radio show. I must also admit that Coasttocoastam.com is on my favorite web site listings and I visit the site often.

Not surprisingly the pilot who photographed the images later stated that he was wrong and that the images were in fact flares being used by the Air Force for night training.

I have no doubt that the pilot may believe the images were flares however the exact same globes have been seen around the world for many years and the odds of obtaining two very distinct similar enhanced images of a pilot in two different "flares" are beyond calculation.

Some have gone so far as to post on blogs that I saw what I wanted to see or that I made up the images. Unfortunately those people did not even read the interview with the news source of they would have read that I did not believe in UFOs and that the objects could have been aircraft developed by the US Air Force. I should also point out that my original intent in investigating the images was to prove to myself that they were not UFOs.

It is not unusual or unheard of for our government to attempt to discredit people regarding UFO events using all sources at their disposal.

Using your favorite image search engine, you can view the color images yourself and make up your own mind as to what they are.

I do know that after taking the images and enhancing them, my views on UFOs has changed drastically in that I now believe that many of the colored lights seen around in the world are some type of aircraft based on my experience.

As more and more people use digital cameras and the resolution continues to increase on those cameras we should all expect that soon there will be a sharp clear image of an actual UFO. We should also expect that many including our government will also most likely claim the UFO image is a fake.

Many have questioned why UFO sightings and stories have seemed to decrease in the past few years. It is my belief that whoever is responsible for the UFOs is aware of new digital camera technology and that flights are now done more carefully and in more remote areas than they were before such technology existed.

I carry a fully charged high-resolution digital camera with me at all times just in case I see a UFO. With a clear up close image and image enhancing technology and software we may be surprised at what will be revealed.

The Black Helicopters.

By now almost everyone has heard of the famous black helicopters used by our government to investigate UFO incidents and to transport officials to persuade witnesses they did not see what they thought they saw.

I should point out that I thought the famous black helicopters were a myth found only in the imagination of UFO conspiracy experts.

My opinion regarding the black helicopters changed late on a Saturday afternoon as the sun was starting to set.

I was leaving a small non-commercial country airport and approaching my parked car when I heard the sound of multiple helicopters approaching. Since I am an aviation enthusiast I recognized the helicopters are being large and most likely military.

As I looked into the distance I saw two large CH47 Chinook helicopters flying in formation and on final approach to the airport runway.

The CH47 is the workhorse of the US military and is used by all branches of service. The CH47 is unique in that is has two rotors. The helicopter has a rotor at the front and a rotor at the rear along with the engines. Most people have seen a CH47 either flying or on news videos as they have been and are still the main aircraft used for troop support in combat areas. The CH47 can best be described as a huge fat banana, with round windows on each side, a big glass cockpit with a nose like a VW Beetle, and two giant rotors. The CH47 is very powerful and very fast for a helicopter and is also used for heavy lifting of equipment and weapons.

The CH47's continued to approach as if to land on the runway but instead both circled the area several times as if they were searching for something.

Eventually the two helicopters landed side-by-side and taxied to a parking area.

I don't think the crew knew I was watching as I inside my car and the car was parked beside a large hanger near the middle of the runway. I did have a clear view of the helicopters.

After a minute or two the back loading ramp came down on one helicopter and several crewmembers exited. Both helicopters continued to idle their engines as if prepared for a quick takeoff.

I noticed the helicopters were not the usual olive drab of the Army or light gray of the other branches of service. These particular helicopters were all flat black and had dark tinted windows all around.

There were no markings on the helicopters of any kind to indicate registration numbers, aircraft inventory numbers, or branch of service. I have never before seen an aircraft with no identification of any kind visible.

The crewmembers seemed to be inspecting something on one of the aircraft as if there was a mechanical issue. The crewmembers were wearing black flight suits with no insignia, black helmets, and the helmet visors were darkly tinted to the point of looking black. The helicopter crewmembers never lifted their visors, which I found strange.

A man, his wife, and a young girl were driving past the airport and apparently they wanted to see the large helicopters. As the man approached the gate to the area where the helicopters were located another crew member came out of the back of the helicopter and used his arms to forcefully warn the family not to open the gate or try to approach either helicopter and that crew member also had a small automatic weapon slung over his shoulder. The family man tried to make conversation with one of the closet crewmembers but the crewmember just faced him with his helmet visor down and said nothing and pointed his finger towards the road as if to advise them to leave.

The family seemed to be surprised and frightened and quickly left the airport.

At this time I was the only person at the airport since it is an uncontrolled airport and has no air traffic control tower or staff. There was no incoming traffic, which is not unusual for late in the afternoon at that airport with darkness approaching. Most flights in and out of that airport are VFR, which means the pilots use their eyes to see where they are flying instead of instruments. Aircraft flying after dark usually land at the larger commercial airport.

I stayed in my position and watched as one crewmember looked around again as if to see if anyone was at the airport. Apparently convinced that no one was watching several more men in identical flight suits and helmets exited the open helicopter. The men gathered and had some type of conversation. After a very short time all crewmembers entered the aircraft, the rear ramp was closed and the helicopters remained on the ground at idle.

I waited to see what has going to happen.

Twenty minutes later it was dark and at that time both helicopters took off in formation into the darkness. Neither helicopter had any type of navigation lights and the helicopters were almost totally invisible in the darkness.

According the reliable sources, known US Government agencies have enough aircraft to make them together one of the largest air forces in the world and that excludes military aircraft. Those figures do not include suspected secret aircraft or secret agencies.

I am now convinced that the black helicopters do exist. I do not know whom the helicopters belong to or what they are used for but I do know they are real.

The next time someone speaks of black helicopters around a particular location I know they are probably telling the truth.

The Strange People.

When some people see something strange sometimes they tell themselves they have not seen what they thought they saw. Other people will have the courage to tell their friends what they have seen.

During certain random time periods many people claim to have witnessed something very strange and you may do the same after reading this story.

Several people contacted me. The sources for this story did not know each other. The people who contacted me all told me the same story. It seems during random time periods there are strange people among us who do not look like they belong.

On many occasions the sources noted they saw many strange acting people on multiple occasions on the same day. The people always wore sunglasses and seemed to have difficulty breathing as if they had been running for a long period of time.

I must admit I was skeptical but since I never dismiss such things before investigating them I said I would try to verify the claims.

Not even two days after being advised of the strange people I began to notice those people in public.

The strange people were just as described to me in that they always wore sunglasses and would always seem to be having difficulty breathing and their mouths would always be wide open. The best way to describe their breathing would be that they look like a fish out of water struggling to breathe.

The strange people were working as delivery drivers, bus drivers, shoppers, police, and most every other type of person among us but they all looked similar and all had the same characteristics and only males have been spotted.

The strange people even all wore the same type of sunglasses that were extremely dark and wrapped around their faces. It appears the strange people seem to all come out and be seen at the same time and apparently in large numbers in that in a 5 mile trip through town in the same day I counted 12 of them.

There is no pattern as to the time of year the strange people are seen. They can be seen during all of the seasons but as far as anyone knows they are only seen during the day.

Normally I would find this story too incredible to believe but I have seen it for myself.

If you are stopped in traffic and notice someone in sunglasses having difficulty breathing and never closing their mouth, odds are you will see many more that same day.

The strange people do not seem to be in need of medical attention, they act as if they have just run a marathon. Pay close attention to details and you will see the strange people seem to be connected somehow and that they look like they do not belong among the rest of us. You will also notice they all look very similar having the same complexion, skin color, hair color, and facial features.

As to who or what these people are there can be many theories or explanations including it is just one big coincidence.

I must also stress that at no time has anyone said the strange people posed a threat to anyone. There is absolutely no reason to be frightened of the strange people.

I would not be surprised if many reading this book have seen the strange people but never told anyone because they thought no one would believe them. If you have not seen the strange people you will now that you know how to identify them.

I know this story sounds like something out of a science fiction movie and crazy but it is true.

A Word From The Author.

It is my hope that by reading this book many people will realize there are things in our world that we may not understand.

Too often people are closed minded to unusual events or incidents and dismiss reports of those events. Our educational system is based on teaching the status quo and not teaching people how to think for themselves or explore new explanations.

Not all cases or stories were included in this book. Some stories were short and simple but true.

Examples of other cases include:

1. The out of control ambulance that crossed the centerline and was about to collide head-on with an automobile when a strange force gently pushed the ambulance back into the proper lane. The ambulance driver, the paramedic next to him, and the occupants of the other vehicle have verified the event did in fact happen and no one has an explanation as to what happened.

2. The individual that can communicate with vehicles by touching them. The individual can accurately state if the vehicle or aircraft has or has not ever been in an accident, if anyone was injured, or if anyone was killed in that vehicle. This person also finds it useful for selecting good used cars in that they can also feel if a car is good or bad and used cars this person has purchased have always been almost trouble free for many years.

3. The man and wife that take turns cooking and each always cooks what the other one is thinking they would like to have for dinner without discussing it in advance.

4. The man who is visited in his dreams by someone he does not know that gives him advice on business and personal decisions that always turn out to be the best decision with the best possible results.

5. A person who can never enjoy a movie because they always know what is going to happen and how the movie will end even if they have never seen or heard about the movie before.

There were other cases that proved to be normal and were not within the supernatural or paranormal and of course they were not included in this book.

If this book has done nothing else for you I hope it has opened your mind. Too many people these days do not investigate situations or think for themselves but instead rely on the media or government officials to explain events to them.

The next time someone tells a story involving supernatural, paranormal, or UFO events I would hope you will take the time to listen to them before you dismiss them as a "nut" or making up the story. Rather than assume the person is making the story up assume they are telling the truth and go from there.

All animals have the gift of curiosity and the need to investigate. We can see investigative abilities in our pets as they confront something new and then try to figure out what they are seeing.

It is natural for all of us to want to investigate supernatural or paranormal events rather than to run away from them or dismiss them as fake.

We are all fortunate in that this is an amazing world we live in full of mystery and unexplained events and our world would be very dull without them.

Supernatural and paranormal events are also a way we are shown there is life after death and other worlds that we do not fully understand.

Do not listen to those who suppress their natural gift of exploration, intuition, and curiosity.

Good luck, and keep an open mind.

With kindest regards,

Mark Kirby

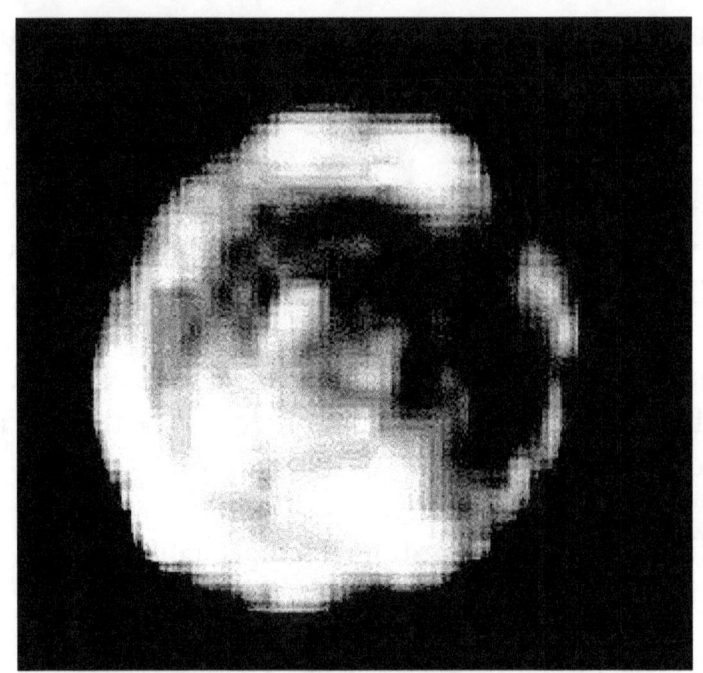

Side view pilot in glowing ball UFO

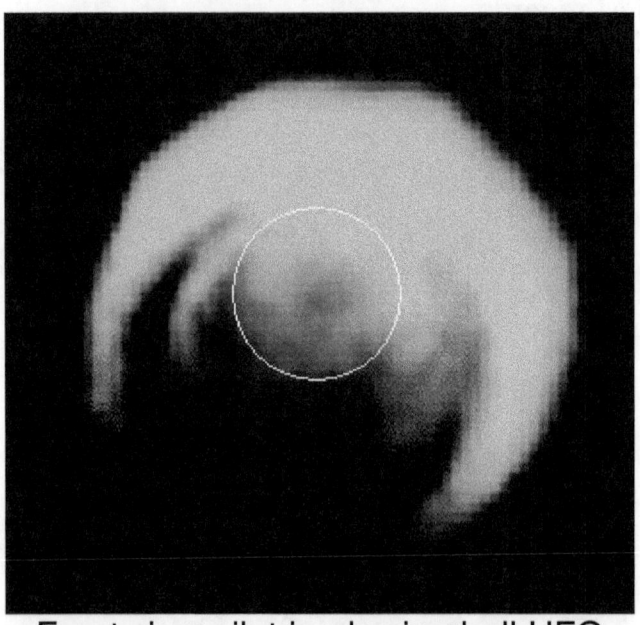

Front view pilot in glowing ball UFO

www.ingramcontent.com/pod-product-compliance
Lightning Source LLC
Chambersburg PA
CBHW051541170526
45165CB00002B/836